Scallop and queen fisheries in the British Isles

James Mason BSc, PhD, MIBiol

Fishing News Books Ltd
Farnham · Surrey · England

This is a Buckland Foundation Book,
one of a series providing a permanent
record of annual lectures maintained
by a bequest of the late Frank Buckland

British Library CIP data
Mason, James *1929–*

Scallop and queen fisheries in the British Isles
1. Queen Scallop fisheries–Great Britain
2. Scallop fisheries–Great Britain
I. Title
338.3′72′411 SH373

ISBN 0 85238 128 X

Published by
Fishing News Books Ltd
1 Long Garden Walk,
Farnham, Surrey, England

Typeset by Paddockglade Ltd, Godalming, Surrey.
Printed in Great Britain by Page Bros (Norwich) Ltd
Norwich, Norfolk

Contents

Figures

Plates

Tables

Preface

The eminent Victorian naturalist Frank Buckland, who died in 1880, left to the nation a sum of money with which to endow a Professorship of Economic Fish Culture. The Trustees of the Buckland Foundation appoint a Buckland Professor annually to review particular aspects of fisheries and to disseminate the available information by means of a series of lectures given at appropriate places in the United Kingdom or Ireland. The lectures, which are subsequently published in book form, reflect Buckland's strong bias towards practical application of knowledge.

The first series was given in Hull and Grimsby in 1930 by Professor Walter Garstang. The series which formed the basis of this book on scallops and queens was given in 1980 in Lerwick, Campbeltown, London, Plymouth and Port Erin. To commemorate the centenary of Buckland's death and the Golden Jubilee of the inaugural series of lectures a Buckland Medal was struck and I had the honour to be the first recipient.

Economic fish culture has been interpreted broadly. That shellfish should have been chosen from time to time as subjects of the lectures is entirely appropriate in view of Buckland's known great interest in, for example, oysters, crabs and lobsters, which were all commercially important in his lifetime. Scallops and queens have assumed great importance only since the second world war, and the development of important fisheries for them has been accompanied by a remarkable increase of interest in them and of available scientific knowledge about them.

The pectinids are a fascinating group culturally as well as biologically. Their attractive shells figure prominently in man's cultural and religious history. The ability of some of their members to swim gives them an added fascination. The book deals with these aspects as well as the development of the scallop and queen fisheries and the coming of processing, which facilitated their expansion. Improved knowledge of their biology and behaviour has enabled us to interpret the catch data and assess the states of exploited stocks. As fishing has expanded, attention has turned to the protection of stocks and recently to the exciting possibility of farming pectinids or replenishing natural stocks. The book ranges widely over all these aspects, representing a timely summing up of the present state of our knowledge of a fascinating group of animals and their commercial exploitation and pointing the way forward. It contains much information of interest to anyone connected with the shellfish industry, be they fisherman, processor, merchant, scientist or administrator.

Much of the knowledge incorporated in the book has been acquired over many years of contacts with friends and colleagues in the shellfish industry and shellfish research. I owe a deep debt of gratitude to them all. I am particularly grateful for the help of a number of them in the preparation and provision of illustrations. Most of the original line drawings were prepared by Mr Robert A Irving, and Mrs M Gammie and Mr A Rice also gave invaluable help. Mr T McInnes helped with the preparation of photographic prints. The use of other illustrative material is acknowledged in the appropriate places. Cambridge University Press kindly allowed me to use much of the material on the age, growth and breeding of scallops which first appeared in my papers in the Journal of the Marine Biological Association of the United Kingdom.

Finally, I am grateful to the Trustees of the Buckland Foundation for inviting me to be the Buckland Professor for 1980 and for their helpful comments on the typescript of this book.

James Mason

1
Introduction

The scallop in history

Bivalve molluscs of the family Pectinidae, the scallops, of which there are some 360 known living species, have long been known to man and have played a prominent part in his history and culture. The Roman naturalist Pliny and other early writers used the Latin word 'pecten', meaning a comb, for scallop shells. Linnaeus first lumped the scallops in with the oysters and the 'great scallop', or simply 'scallop', of Western Europe (*Plate 1*) was known as *Ostrea maxima*. In 1776 Müller recognized the need to distinguish between them, and following Pliny he named the scallops *Pecten*, and *Ostrea maxima* became *Pecten maximus*, the type species. Since Müller's time the genus *Pecten* has been subdivided and many species, obviously scallops but sufficiently different from *P maximus*, have acquired other names. One such is the other subject of this book, the queen, which is known as *Chlamys opercularis* (*Plate 2*).

Pectinids are among the most attractive of all bivalves, both in shape and in the range of their colours, and they have caught man's imagination with the result that they have been used throughout the world as ornaments and emblems, as well as food, from prehistoric times to the present day. The shells of both the scallop and the queen (*Fig 1*) are almost circular, and the dorsal hinge line, which is almost straight, is extended into auricles or ears. In the scallop the auricles are almost symmetrical, but in the queen there is a marked asymmetry. In the scallop, the white lower (anatomically right) valve, on which it lies, is deeply convex, while the upper valve, generally red or brown, often marbled, is almost flat. In the queen both valves are convex, the upper, or left, valve being slightly more so than the right, and the colour varies from a delicate yellow or pink, through red to brown.

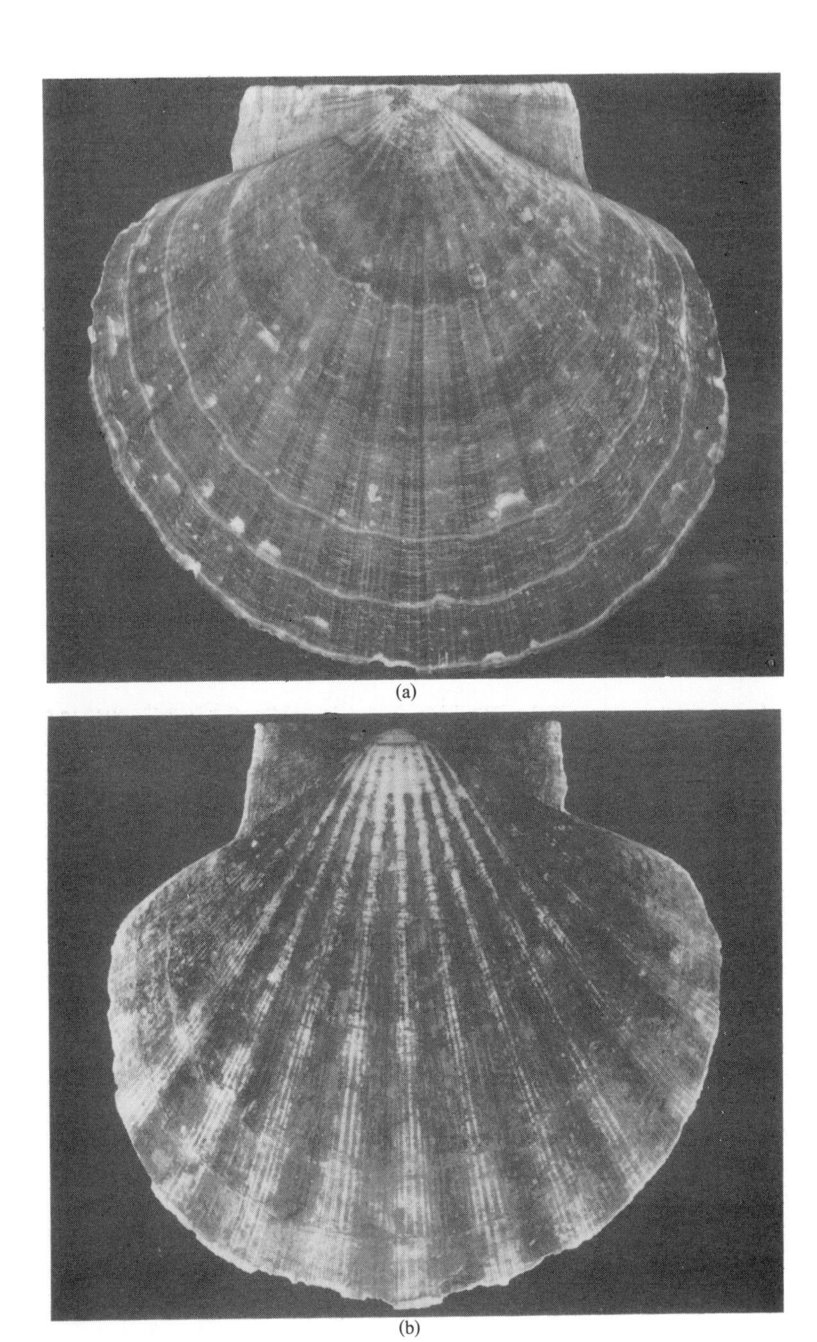

(a)

(b)

Plate 1 The shell of the scallop, *Pecten maximus*, (a) upper left valve and (b) lower right valve, showing growth rings

(a)

(b)

Plate 2 The shell of the queen, *Chlamys opercularis*, (a) upper left valve and (b) lower right valve, showing growth rings and the byssal notch

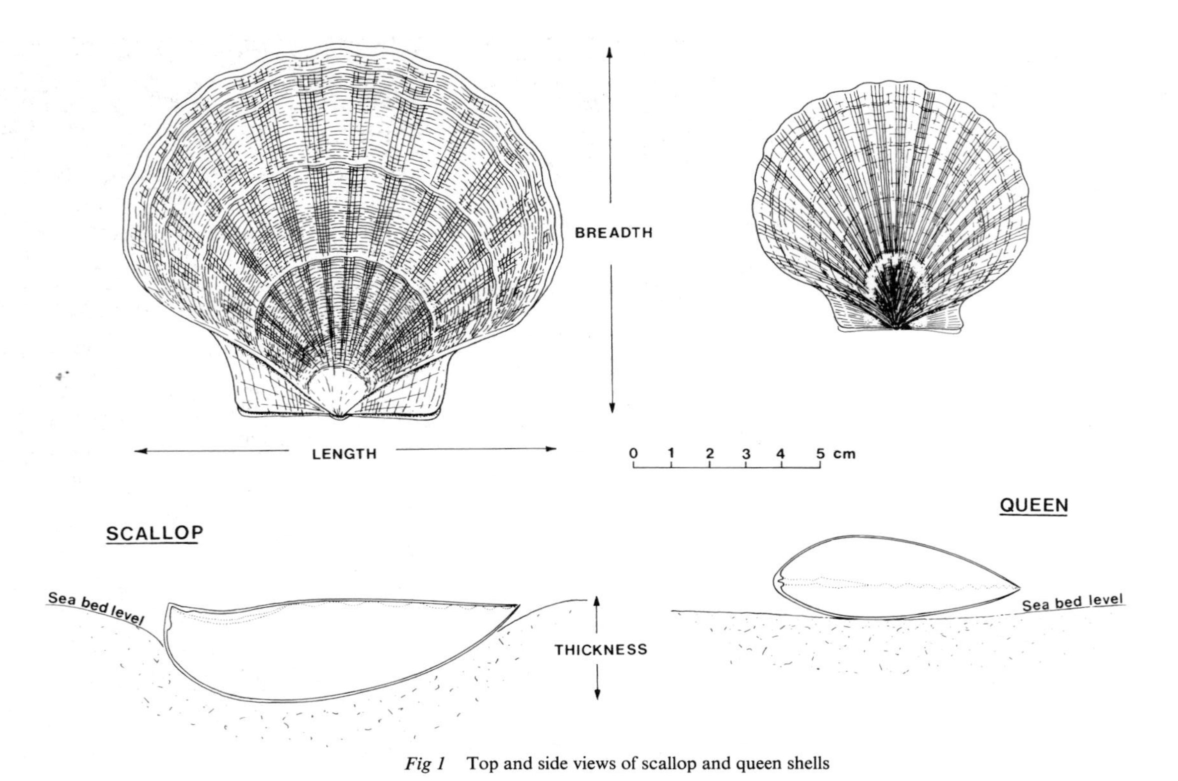

Fig 1 Top and side views of scallop and queen shells

In both species the valves are plicated or ribbed. The ribs radiate out from the umbonal region and interlock with those of the other valve where their edges meet. The number of ribs remains constant throughout life. There are more ribs on the queen (14–22, commonly 19–22) than on the scallop (10–18, commonly 15–17) shell. The outer surface of the shell bears prominent concentric markings, or striae, which in places are crowded together to form pronounced rings, usually white, which are formed annually, and can be used to age the animal.

The two valves of the scallop and queen shell are held together along the hinge line by an external ligament. Inside there is an elastic ligament, triangular in section, which when naturally expanded causes the valves to gape. To close the shell valves, the hinge is compressed by the action of the adductor muscle, which connects the valves internally. The inner surfaces of both valves are completely lined by the delicate mantle, which secretes the shell, and when the animal is at rest (*Plate 3*) there can be seen many long, delicate flesh-coloured sensory tentacles round the edge of the mantle, constantly being extended and retracted, testing the water. Among the bases of these tentacles are numerous tiny, elaborate eyes, which glisten like tiny green jewels. Behind these are the deep inner mantle folds which meet loosely to form a curtain, the edges of which are lined with shorter tentacles. These folds, the frill, flesh-coloured and mottled with brown or green, are constantly parting and closing.

Plate 3 The queen at rest, showing the mantle curtain, tentacles and eyes

As shown in *Fig 2*, if the right shell valve and its contiguous mantle are cut away the soft body parts also present a colourful appearance. In the centre is the white adductor muscle, which linked the valves, but which has now been cut through. Anterior to it are the pale brown kidney and the hermaphrodite gonad, the distal female part bright red from the ripe and developing eggs, the proximal part cream coloured and containing the male sexual cells. Dorsally the mouth, surrounded by palps, leads into the greenish-black digestive gland, from which the rectum leads down the posterior edge of the muscle to the anus. Ventral to the muscle are the folded, orange-coloured gills.

Small wonder, then, that such attractive animals have long excited man's attention. The scallop shell was associated with the birth of Aphrodite (Venus) in Greek and Roman mythology, the earliest evidence being a Greek urn, found on the north shore of the Black Sea and dating from *ca* 400 BC. This, like the best known example, namely Botticelli's painting *The Birth of Venus*, depicted Venus rising fully formed and mature from an open scallop shell. The form of the shell was also used in a number of ways in ancient Greek and Roman civilisations, appearing on burial urns, gravestones and coffins. Domestic utensils from the third and fourth centuries AD, including flasks, vessels, lamps and ladles, took the shape of scallop shells. The shell appears on coins from the second and first centuries BC from Italy and Spain. Roman wall and floor decorations took the form of scallop shells – the best known are at Pompeii and Herculaneum, though a mosaic floor dating from *ca* 150 AD has been uncovered at Verulamium (St Albans) in England. These instances from Greek and Roman civilisations are probably *Pecten jacobaeus,* a Mediterranean form which is similar to, but smaller than, *P maximus.*

Perhaps the best known association of the scallop shell is with the shrine of St James the Great at Compostella in northwest Spain. The story has it that, after James was beheaded by Herod Agrippa, his remains were rescued by his disciples and taken away in a boat, which came ashore at Iria, a place owned by a pagan lady, Lupa. After a series of miracles, Lupa was converted to Christianity and allowed the disciples to bury James in her palace. In the eighth or ninth century AD a church was built on the site, which became known as Santiago di Compostella. The shrine acquired first local and then more widespread fame, and

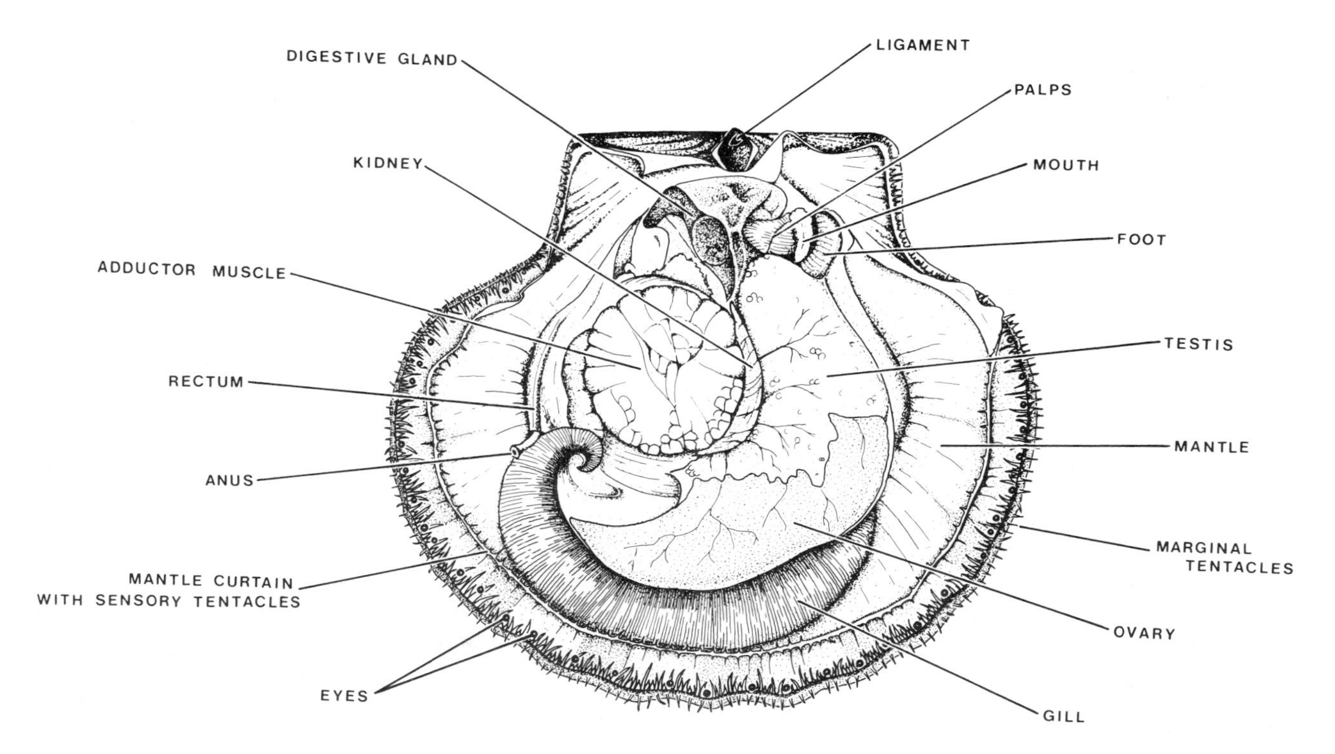

Fig 2 Scallop with right shell valve and mantle lobe removed to show viscera

now, after Jerusalem and Rome, it is the subject of the third greatest pilgrimage in Christendom and its badge of pilgrimage is the scallop shell, which the pilgrims wear on hat or cloak.

This association possibly arose from the miraculous rescue by St James of a horseman from drowning – on emerging from the waves both he and his horse were said to be covered with scallop shells. A more likely explanation is that scallops, which were brought to Compostella from the sea side some 16 miles away for food, were recognized by an enterprising hawker as a potential badge because of the attractive shape and since the scallop, like St James, had come from the sea. Whatever the origin, the scallop shell was the emblem of this pilgrimage by the twelfth century, and miracles, such as the cure of goitre, were attributed to the shell. Stories of the personal intervention of St James to bring victories to Christian armies over the infidel led to James becoming the patron saint of Spain.

Certainly promoters of the Santiago pilgrimage have found the scallop shell an effective symbol, possibly because of its attractive design. It has been conjectured that Scotland, which claims the bones of another apostle at St Andrews, might have become the centre of a pilgrimage if the saltire had the same subtlety of design as the scallop shell.

Though *Pecten jacobaeus* bears the name of pilgrim's scallop, or St James's scallop, this is a Mediterranean species and is not found off northwest Spain. As Rees (1957) says, the true pilgrim's shell is *P maximus*, which is abundant there.

The scallop shell commonly appears in coats of arms, and has done so for some 700–800 years. It almost always appears hinge uppermost, not because this is correct zoologically, but because this is how it was worn by pilgrims or hung at home by them. It appears in the arms of families who have made pilgrimages to the Holy Land or other shrines, or who have made long voyages, gained great victories or held important naval commands (Lovell, 1884; Bellew, 1957). Perhaps the use of a scallop shell as an emblem most commonly in the public eye today is as the badge of the Shell Oil Company, though there it is inverted (See Cox, 1957).

The religious significance of scallop shells has extended to their use in baptismal services, a practice which was first noted in the fifteenth century and still persists in some churches, though nowadays the vessels are usually silver gilt copies. My own younger daughter was baptised with water from the right valve of a real *Pecten maximus*.

On a more mundane level, scallop shells, the flesh probably having been taken for food, have been used down the centuries in many parts of Europe as cups and dishes, and by fishermen as primitive lamps in their huts. However, not being shore-dwellers, live scallops were probably inaccessible to prehistoric man, which would account for their relative scarcity in kitchen middens. Such shells as have been found, for example in the Obanian culture, often show signs of wear from being used as utensils (Lacaille, 1954), and had probably been washed up on the shore as empty shells (P Mellars, personal communication).

Distribution and habitat

The geographical range of *Pecten maximus* along the European Atlantic coast extends from northern Norway south to the Iberian peninsula (Tebble, 1966), and it has been reported off West Africa, the Azores, Canary Islands and Madeira. According to various authorities it occurs from just below low water mark (Tebble, 1966) to a depth of 100fm (183m) (Forbes and Hanley, 1853), but it is commonest in 10–25fm (18–46m). It is common round the coasts of the British Isles, preferring bottoms of clean, firm sand, fine gravel or sandy gravel, sometimes with an admixture of mud.

Chlamys opercularis is distributed from northern Norway and the Faroes south to the Iberian peninsula, the Azores and the Canary Islands, extending into the Mediterranean and Adriatic (Tebble, 1966; Broom, 1976). It occurs in depths down to 100fm (183m), but is again commonest in depths less than 25fm (46m). It is found on the same areas of sea bed as the scallop, though it can also live on harder gravel and shelly bottoms because, unlike the scallop, it does not recess into the sea bed.

Owing to the availability of suitable substrata both species are essentially coastal. Both tend to be most abundant just inside or just away from areas of strong currents, *eg* the Calf Sound (Isle of Man) and Kilbrannan Sound (Clyde Sea Area). The distribution of both species is patchy (Baird and Gibson, 1956; Rolfe, 1973; Soemodihardjo, 1974), patches sometimes being predominantly of one age group, suggesting that the larvae may be gregarious.

Queens are generally much more numerous than scallops. Baird and Gibson (1956) found densities of scallops as high as one per 1.65m^2, and one scallop per 5 or 10m^2 is considered a commercially fishable concentration (Mason and Colman, 1955; Gruffydd, 1972). I have found queens in layers three or four thick on the sea bed, at a density of 500–1000 per m^2, though this is exceptional. Between such dense patches of either species are areas which appear capable of supporting them but in which they are scarce or absent, presumably because they happen not to have settled there.

2
History of the fisheries

Although both *Pecten maximus* and *Chlamys opercularis* have been fished round the British Isles (*Figs 3, 4*) for several centuries, certainly since the sixteenth century (Stanley, 1967), they did not assume great commercial importance until after the second world war.

Scallop fishing is reported on the west coast of Ireland as early as the sixteenth or seventeenth century (Duff, 1976), though this was almost certainly for home consumption and not commercial exploitation. In the mid-nineteenth century scallops caught there by small boats using an oyster dredge were recorded as being used as bait on longlines and for making soup. By the end of the nineteenth century Irish scallops were being marketed in London. Landings in the Irish Republic remained low during the 1930s, increased during the early 1940s, and fell away after the second world war to 74.4t in 1950, thereafter fluctuating between 64.9 and 215.3t per year in the 1950s (*Fig 5*).

It is difficult to obtain accurate data of landings in the official British sea fisheries statistical tables before the early 1930s, since prior to that landings of scallops and queens were so small that they were often included with other shellfish. Scallop and queen data were combined until 1970 in Scotland and until 1972 in England and Wales, but it has proved possible to separate them by recourse to raw data.

Landings of scallops have been made in England and Wales for over a century. Up to the early years of the twentieth century much of the catch, some of which was sent to Billingsgate Market, London, and some used as bait, was taken on the east and south coasts of England (Franklin, Pickett and Connor, 1980). In the 1920s and 1930s the landings were mostly at Emsworth (Hampshire) and Brightlingsea (Baird, 1952), but landings had declined

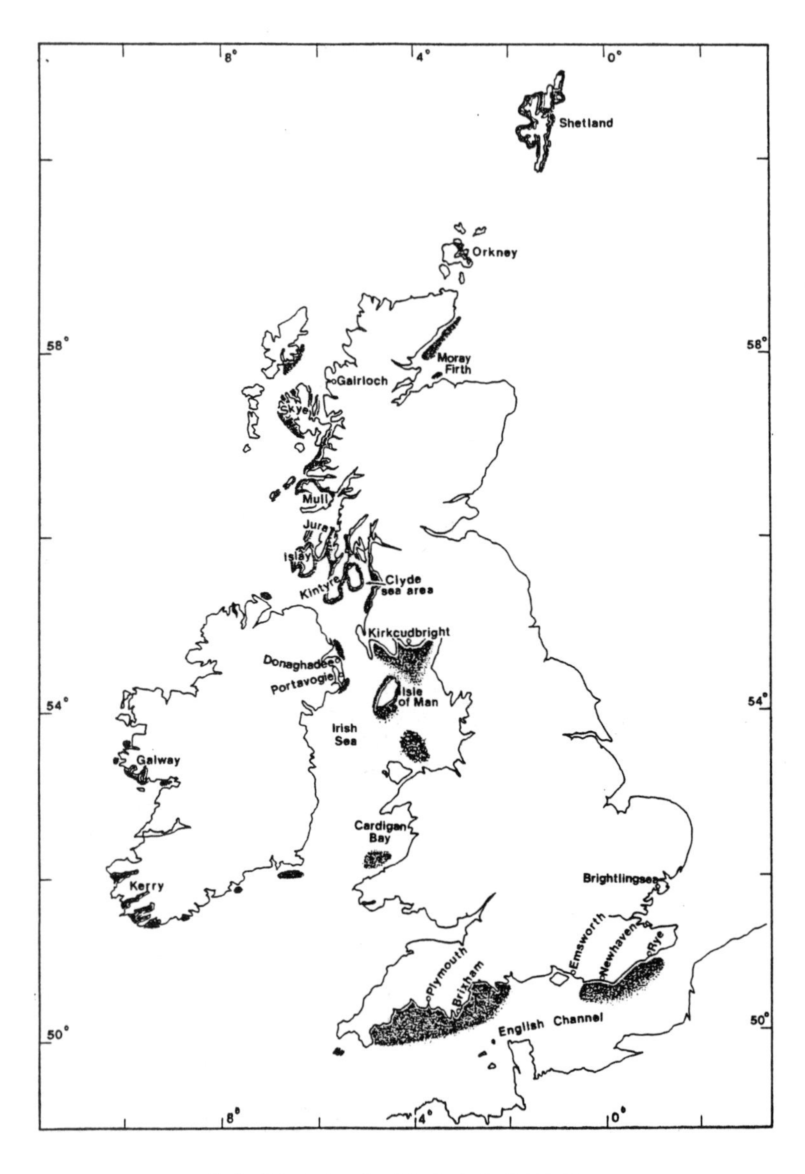

Fig 3 Principal scallop fisheries in the British Isles

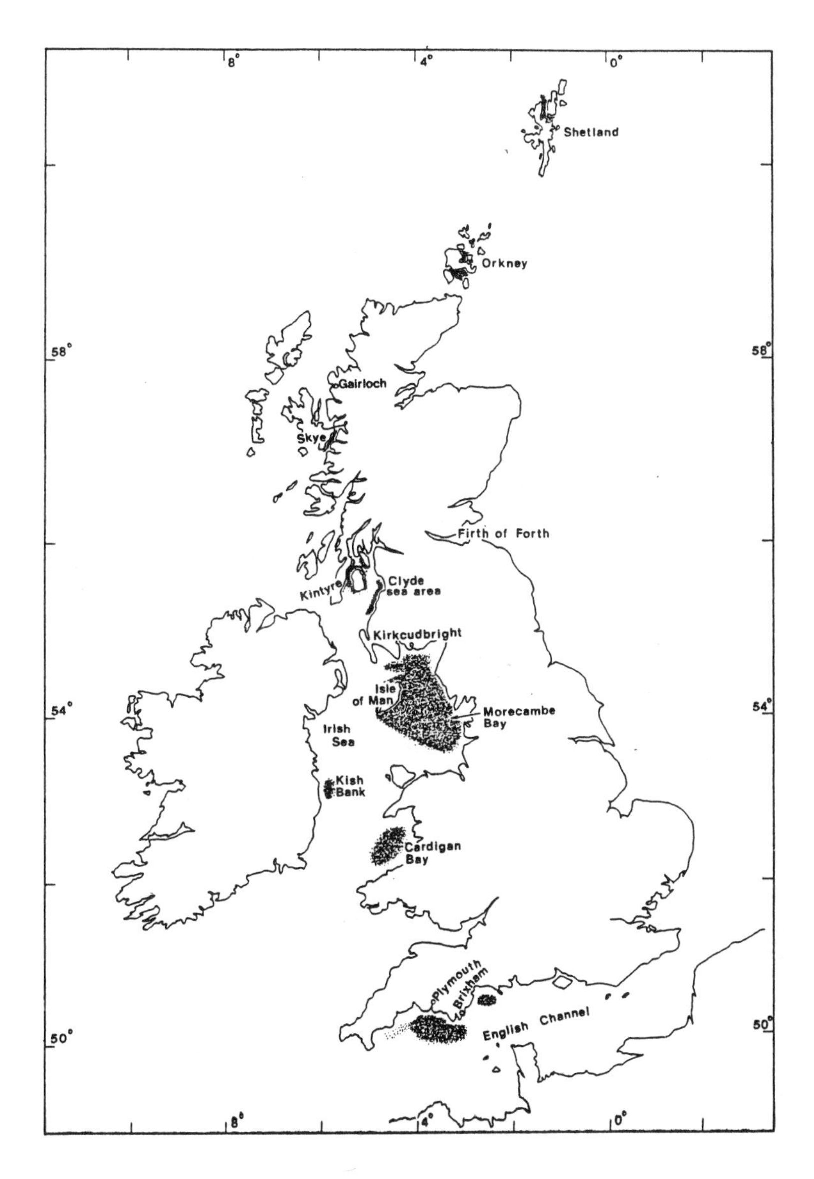

Fig 4 Principal queen fisheries in the British Isles

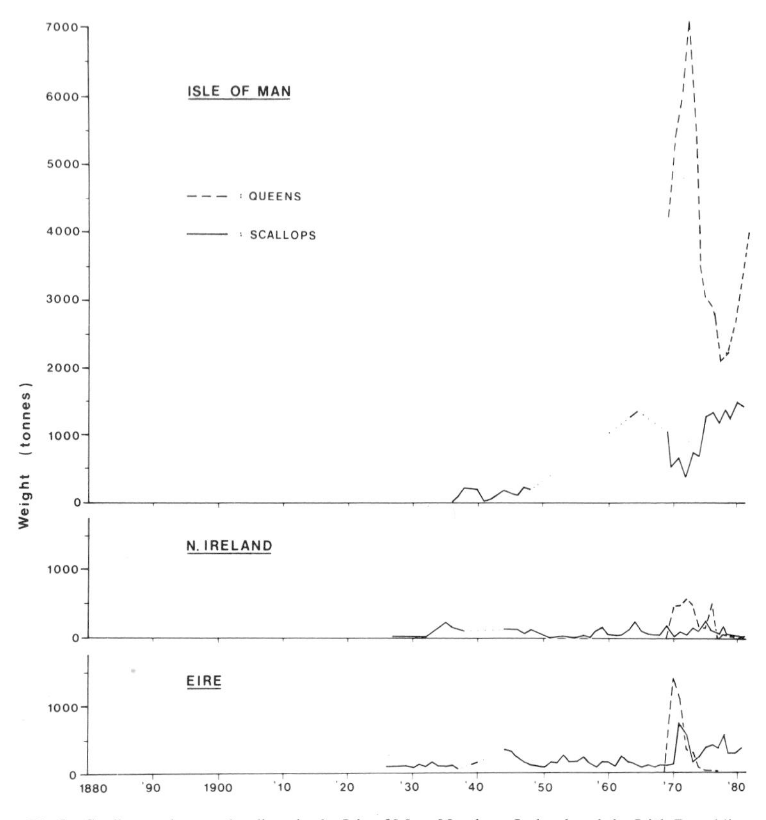

Fig 5 Scallop and queen landings in the Isle of Man, Northern Ireland and the Irish Republic

almost to nothing by the early years of the second world war, and remained low for many years (*Fig 6*). The decline was probably due to the low price, which failed to attract new blood to the fishery (Baird, 1952).

Landings in Northern Ireland were low in the late 1920s but a commercial fishery became established in 1933, and landings up to 1938 averaged 137t. The fishery provided winter employment for up to 90 boats. No data are available for the war years. There was a steady decline to 6.7t in 1955 but a recovery in the late 1950s (*Fig 5*). Scallop fishing round the Isle of Man started in 1937 when an Irish fisherman drew the attention of Manx fishermen to the possibility of exploiting local stocks of *Pecten maximus,* and soon supplanted herring as the most important Manx fishery. Landings reached 228.5t in 1947 and the value

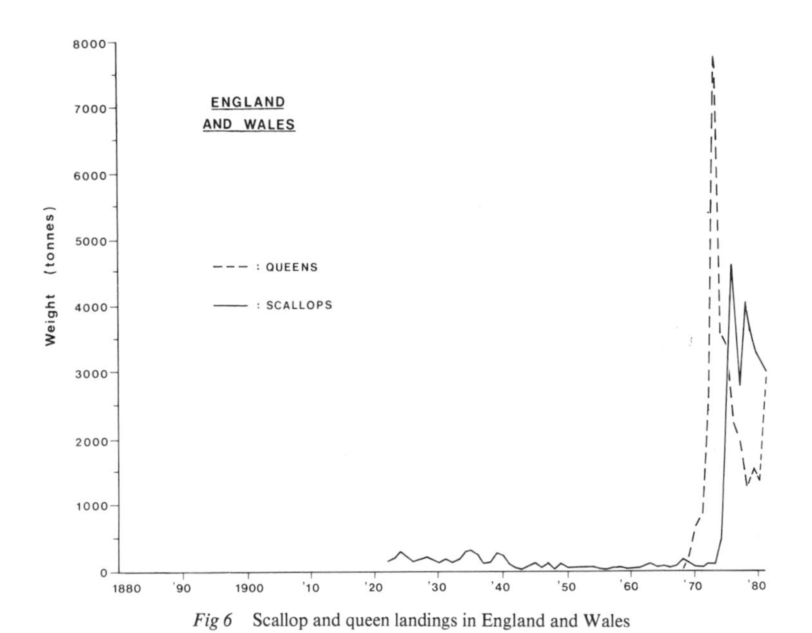

Fig 6 Scallop and queen landings in England and Wales

exceeded £11,000. Up to 13 boats were then involved, mostly small boats in the size range 25–49ft (7.6–14.9m), fishing off Port Erin and Peel on the west coast.

About 1880 a dredge fishery for the queen developed in Scotland. The catch, almost entirely from the Firth of Forth (*Fig 4*), was used as bait in the then extensive small-line fishery. Landings reached a peak of 1,449t (value £3,300) in 1891, remained high throughout the 1890s, but declined gradually as small-line fishing declined, and had virtually ceased by the early 1920s (*Fig 7*), stopping altogether in the early 1930s. A dredge fishery for the scallop started in the Clyde Sea Area in the 1930s, about the same time as the Manx fishery. The fishery remained small during the 1940s and 1950s, landings averaging 111.8t (value £8,200), and provided a useful seasonal income. Fishing was carried on only during the colder part of the year, chiefly October–March, and was adopted to fill in time during the off-season of another fishery or because catches had not been too good in another fishery. The number of boats was small, seldom exceeding ten.

The catch of scallops from all parts of the British Isles was sent live in the shell to Billingsgate Fish Market, London, being con-

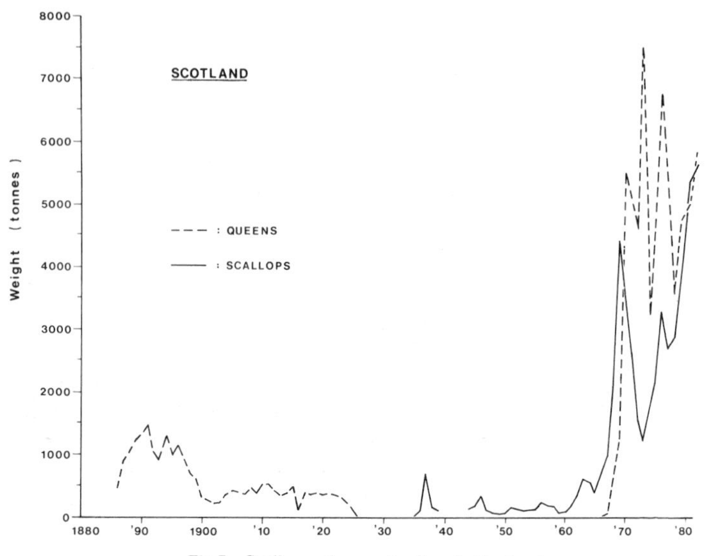

Fig 7 Scallop and queen landings in Scotland

signed in sacks, with the scallops carefully packed, round-valve downwards in order to retain as much water as possible. Fishing had to stop in the spring, for two reasons: firstly, because the scallop, after being in excellent condition throughout the winter and early spring as the roe develops and being at its best in the early spring, then spawns (see *Chapter 7*), the roe becomes empty and the value falls; and secondly, because of the difficulty of keeping the scallops, sent alive in the shell, in good condition during the long journey to Billingsgate in the warm weather. Scallops and queens, unlike the mussel and cockle, do not have the two shell valves meeting completely all round the margin and so they rapidly become dehydrated.

This then, was the picture up to the end of the 1950s. The 1960s and 1970s saw a radical change in the fishery. In addition to the home market, there developed a large demand for scallops on the continent of Europe, and demand far exceeded supply.

The expansion of the scallop fishery started in Scotland in 1961 and 1962, when it was still confined to the Clyde and 10–15 boats were taking part. Attention was then drawn to the fact that the roe rapidly recovers after the spring spawning, is full again by late spring or summer, and remains full until the late summer spawning (Mason, 1972a; 1978). About that time processing factories, where the flesh was extracted, first appeared in Scotland, and it became possible to market frozen meats, many being exported to the continent of Europe, especially France, Belgium and Spain, though there was and still is a demand at Billingsgate for large scallops ($\geqslant$125mm) in the shell. Fishing thus became possible all the year round. From the early 1960s boats from the Clyde began to explore grounds further afield, particularly West of Kintyre (Sound of Jura, Gigha, Islay, *Fig 3*) where research vessels had revealed concentrations of scallops, and they found many beds. The small (up to 30ft, 9m) Clyde boats, fishing two or three 4ft (1.22m) or 6ft (1.83m) dredges were joined by bigger boats, mostly of the Scottish side trawler/seiner type, using more and bigger (5–6ft) dredges. Later, however, the trend towards larger dredges was reversed and dredge size became standardized at 4ft (1.22m), where it remained for many years. From 1980 onwards the practice of using even smaller (2½ft, 0.76m) dredges became widespread.

In 1967 beds off the northwest mainland were first exploited on a commercial scale and the fishery there expanded rapidly. In 1967 a fishery started at Shetland, but this has never reached the size of the west coast fisheries. Dredging has occurred sporadically also at Orkney, and from 1977 a fishery has developed in the Smith Bank area of the Moray Firth.

As the fishery expanded the number and size of boats increased. Larger seine-netters with up to 10–12 dredges became common (*Plate 4*), and more recently 80ft adapted Dutch beam trawlers joined the fleet. Scottish scallop landings reached 4,413t (value £538,000) in 1969, when the fishing fleet had risen to more than 70 boats. They fell during the early 1970s but increased again later to 5,527t (value £3,236,000) in 1981, both weight and value being the highest ever. By then, the number of Scottish boats fishing had risen to 100.

It was known that scallops occurred in fishable concentrations all round the Isle of Man (Mason, 1959a) and the scallop fishery increased there at approximately the same time as in Scotland as processing factories were established on the island. Unfortunately no official landings data are available for the period up to 1968, though it was estimated that annual landings had risen to some 1,300t by 1963–1965 and in 1969 they were 1,045t (value £150,000). The number of boats had reached 50 by the late 1960s

Plate 4 A 65ft (18.3m) Scottish seine-net boat dredging for scallops

and has since remained fairly constant. Most of the boats were still in the size range 35–50ft (11–15m).

Scallop landings in England and Wales remained low throughout the 1950s, 1960s and early 1970s, fluctuating between 14 and 195t. Brixham was at first the most consistent port, though Plymouth became increasingly important from about 1966. Landings of scallops in Northern Ireland fluctuated in the late 1950s and 1960s, reaching a peak of 218t (value £19,600) in 1964 but there was no spectacular increase such as occurred in Scotland and the Isle of Man. Fishing then and subsequently has been mainly by small boats at Portavogie and Donaghadee. In the Irish Republic landings remained at a low level in the 1960s, fluctuating between 58.1 and 216.2t.

In the late 1960s a revolutionary development occurred. While dredging for scallops fishermen often caught queens, which in some areas were so abundant that the dredge rapidly became filled with them. For example, off the Cock of Arran in the Clyde the research vessel *Mara* took 20 baskets of queens (10,000) in one haul of 2 minutes duration using one 6ft (1.83m) dredge

(Mason, 1972a). Research divers confirmed that the queens there lay three or four deep on the sea bed. Although the queen is acceptable for human consumption, little attention had hitherto been paid to it by fishermen, who regarded it as trash. The chief reason was its small size, requiring more effort in shelling than scallops for an equivalent amount of meat. However, local abundance, together with the recent advent of processing factories around Scotland, made queen fishing possible. The discovery of a lucrative market in the United States of America, owing to a declining yield in the bay scallop, *Argopecten irradians* (Lamarck), fishery and consequent high prices, resulted in 1967 in some Scottish boats switching from scallop to queen fishing, more boats joining in, bringing the total fishing scallops and queens in Scotland to more than 80, and a rapid expansion in queen fishing. This started in the Clyde, where scallop catches were at the time poor owing to recent poor recruitment. But it soon spread to other areas, first to Orkney, then the North Irish Sea and Shetland. Fishing has continued in all these areas except Orkney. Scottish landings rose rapidly to a peak of 7,493t (value £704,000) in 1973, fell to 3,233t in 1974 as the market declined, and have subsequently risen and fluctuated, and the highest ever value of £1,302,000 was achieved in 1981.

As the queen landings increased in the late 1960s and early 1970s, so the scallop landings decreased from the 1969 peak to only 1,210t in 1973, but they rose again as queen landings decreased and have since fluctuated around the 3,000t mark. Scottish scallop landings achieved their greatest ever weight (5,527t) and value (£3,236,000) in 1981.

Queen fishing by Manx boats soon followed the Scottish fishery, starting in 1969 when the local scallop beds were said to be scarcely profitable (Paul, 1978), and many boats switched their effort to queens. There are now some 50 boats supplying eight processing factories on the island. Queen landings rose to a peak of 7,128ft in 1972 and subsequently declined until 1977, though the years 1978–1981 have shown a steady recovery. At first most boats fished for queens all the year round, but as demand decreased many fished for scallops in the winter (scallop fishing is restricted by the Isle of Man Government to the period November– May) and queens in the summer. As queen landings increased in the early 1970s scallop landings fell, as low as 360t in 1972, but as queen landings later fell scallop landings rose, and continued to rise even though queen landings recovered in the late 1970s. The highest-ever Manx scallop landings, 1,533t (value £908,000), were achieved in 1980.

Northern Irish boats joined in the queen fishing in the Irish Sea from 1970 and landings reached a peak of 584t (value £47,600) in 1972, but landings declined as marketing trends made whiting and Norway lobster fishing more profitable, and had virtually stopped by 1977. Fishing for queens in the Republic of Ireland started in 1970, when a hitherto unexploited bed was found on Kish Bank off the east coast. Landings in 1970 were 1,367.4t (value £70,000), but they declined there also and since 1974 have been negligible.

Interest in queen fishing was also aroused in England and Wales at about the same time as in the Isle of Man and Ireland. Several English boats joined the Scottish fleet fishing in the North Irish Sea, landing principally at the Scottish Solway Firth ports but also in northwest England. Fishing spread south in the Irish Sea into Cardigan Bay (west Wales) and landings were also made at Welsh ports. In 1969 also stocks of small queens were discovered in the English Channel, and by the end of 1972 most of the boats fishing out of Plymouth and Brixham were fishing for queens. As in Scotland, landings rose rapidly to a peak in 1973, the total for England and Wales being 7,720t (value £450,000), and like Scottish queens almost all were processed for the American market, and they too fell in 1974 as the American market declined. But whereas Scottish queen landings soon recovered, those in England and Wales have continued to fall, the 1978 landings being 1,239t (value £256,000).

This continued decline in queen landings in England and Wales was due to a switch to scallop fishing as a result of development of overseas markets. The result was a spectacular increase in scallop landings from only 83t (value £16,000) in 1973 to 4,580t (value £1,890,000) in 1976. Landings have since fallen a little but have remained above 3,000t. The 3,814t landed in 1978 fetched a record value of £2,277,000 and from 1975 to 1978, for the first time since 1950, scallop landings in England and Wales exceeded those in Scotland. Though there was some fishing in the Irish Sea, the initial expansion occurred almost entirely in the English Channel, in both the western (Plymouth and Brixham) and eastern (Newhaven and Rye) areas. These have remained the main fishing centres but boats have tended to fish further from home as more virgin grounds were discovered, and now scallops are fished along virtually the whole south coast of England. The larger vessels from Brixham, Plymouth and other south coast ports are exploiting still more recently discovered offshore grounds (Franklin, Pickett and Connor, 1980), both in the English Channel and in Cardigan Bay, and are remaining several days at sea.

The English Channel fishing is seasonal, though some boats fish all the year round. In the western channel scallop fishing is mainly in the summer, and many boats switch to mackerel in the winter. In the east channel, on the contrary, scalloping predominates in the winter and some boats switch to trawling for sole in the summer. Up to 180 boats, ranging in size from 25ft (7.6m) to 100ft (30.5m) took part in the fishery in 1978, all, with the exception of a few in the Irish Sea, fishing the Channel beds.

The average price of scallops changed little before the second world war, being £29.47 per tonne in 1923 and still only £24.02 in 1932. During and after the war it increased rapidly. As the demand for scallops and the landings increased, the price increased in excess of what could be accounted for by inflation, reaching £624.75 per tonne in the United Kingdom in 1978, though falling to £528.81 per tonne in 1979 as the demand decreased owing to the continental market becoming flooded with Japanese scallops, *Patinopecten yessoensis* Jay. It recovered in 1980 to £645.11. The price of queens has risen steadily and reached £220.32 per tonne in 1980 (*Fig 8*).

In conclusion it can be said that the recently-expanded scallop fishery and the newly-developed fishery for queens for human consumption have now achieved considerable importance, particularly in the United Kingdom, where the combined value at first sale reached £6.9 million in 1980, and in the Isle of Man, where it reached £1.66 million in 1981.

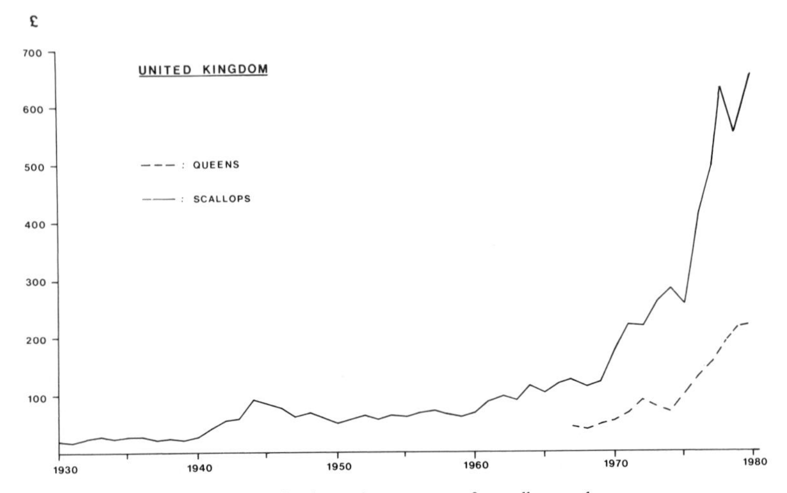

Fig 8 United Kingdom prices per tonne for scallops and queens

3
Methods of capture

Perhaps the simplest and most primitive gear used for catching scallops is the hand-net, which is used in the west of Scotland, Ireland (Gibson, 1957) and the Outer Hebrides. It consists of a long wooden pole to one end of which is lashed a metal ring some 8in (20cm) in diameter carrying a netting bag. In Ireland this net is known as a 'brideog'. It is used only in shallow water to pick up an individual scallop observed from the surface. It can be used only in calm conditions when visibility is good, and is not very efficient owing to the scallop's habit of recessing in the sea bed (see *Chapter 5*).

The traditional method of catching scallops is by dredge. Basically this consists of a rigid metal frame with a netting bag attached, which is towed along the sea bed. Early dredges were probably oyster dredges, either with no blade or with a plain blade set at an angle to the sea bed. The subsequent addition of a toothed bar served two purposes, first to increase the efficiency by digging out the recessed scallops, and second to allow the escape of bottom deposit and trash between the teeth so that the bag filled less rapidly and longer tows were possible.

As recently as the 1920s large, fore-and-aft rigged sailing vessels each towing a dredge 5–6ft (1.5 to 1.8m) wide were used in the English Channel fishery (Davis, 1927) and sailing and rowing boats were used in the Irish fishery (Gibson, 1957). The dredges were shot and hauled by hand and towed from a rope warp.

The Irish south coast dredge (*Plate 5a*) had a rigid rectangular mouth 3½–5½ft (1.1–1.7m) wide with teeth welded to the bottom edge. Three rigid metal rods formed a towing bridle with a ring, arranged so that the mouth was at such an angle that the teeth dug into the sea bed at 45° (Gibson, 1957; Duff, 1976). The dredge used on the Irish west coast is similar to that which became stan-

Plate 5a Irish south coast dredge *(By courtesy of A R Brand)*

dard in fisheries throughout the British Isles. The standard dredge consists of a strong triangular iron or steel frame with the side and inner members extended rearwards and bent downwards (*Plate 5b, Fig 9*). There is a towing eye at the apex of the triangle and a toothed bar or blade is bolted to the bent members so that the teeth dig into the sea bed at an angle of some 60–70° (Mason, 1972a; Chapman, Mason and Kinnear, 1977; Strange, 1977). The scallops are scraped into a bag which usually consists of a belly of linked steel rings secured behind the blade and a back of netting. The netting back extends forward to a crossmember ahead of the blade, and the rear end of back and belly are attached to the tail bar, formerly made of wood but now of tubular steel. In Scottish and Manx scallop dredges the belly rings are commonly of ¼in (6mm) steel and have an internal diameter of 3¼in (83mm). The stretched mesh of the back is usually 3–3¼in (76–83mm), sufficient to allow much deposit and trash to pass through.

In the early days after the second world war the size of dredge depended on the size of boat. The smaller boats, 25–39ft (7.6–11.9m) long, used two or three small dredges, 3½–4ft (1.1–1.2m) wide, towed over the stern, each on its own rope warp. The then large boats, up to 50ft (15.2m) long, towed four 6ft (1.8m) dredges

Plate 5b Standard 6ft (1.8m) dredge

on separate rope or wire warps, two over the stern and one over each quarter. The dredges are shot over the stern the right way up, with the boat steaming ahead and an amount of warp paid out about three times the depth of water. The dredges are towed for 1–1½ hours, at a speed of about 2 knots, and then hauled by means of a powered winch. The teeth are allowed to rest on the boat's rail, which is often reinforced and ridged, and the dredge contents are emptied onto the deck through the mouth by lifting the tail bar.

In the mid 1960s Scottish and Manx dredges became standardized with a width of 4ft (1.2m), 12 teeth 1in (25mm) wide protruding some 2in (51mm) and spaces 3in (76mm) wide between them. The dredges are fished in gangs from a towing bar attached to the warp by means of a bridle, the number of dredges used depending on the size and power of the vessel. The towing bar may or may not have a large rubber wheel at each end. One bar is towed over each quarter and each may have two to five dredges attached to it (*Fig 9*). The gangs of dredges are shot, hauled and emptied as described above for single dredges.

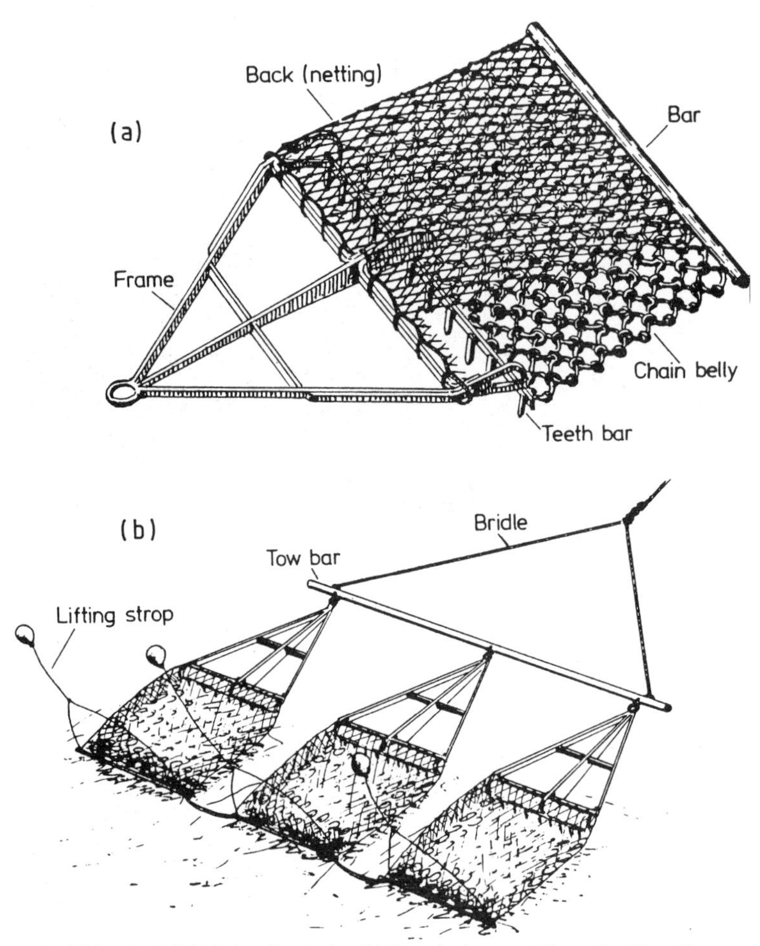

Fig 9 (a) Standard 4ft (1.2m) scallop dredge; (b) Three dredges towed from a bar (*From Strange, 1977*)

In the late 1960s, when the boats were exploiting more and more new beds, and were being forced to fish on rough, stony ground, the Scottish fishermen incurred progressively greater losses of, and damage to, their dredges. This led first to the construction of heavier and more robust dredges and then to the development and patenting of the 4ft dredge with a spring-loaded tooth bar (*Plate 6, Fig 10*) which 'gives' when boulders or other obstacles are encountered, thus reducing damage. Their teeth are narrower and longer than in the standard dredge, *ca* 21mm wide and protruding some 86mm. These dredges had virtually replaced

Plate 6 4ft (1.2m) spring-loaded dredge

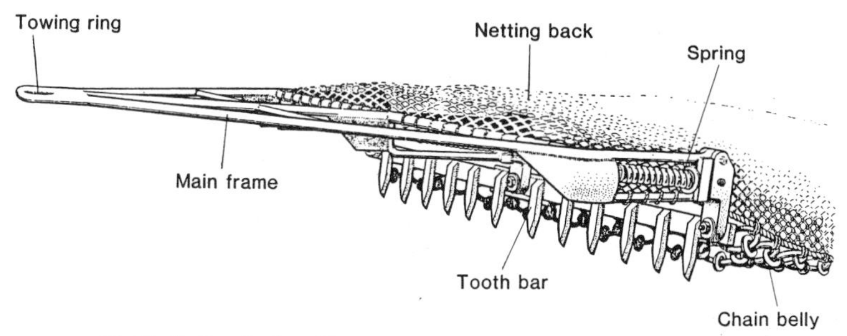

Fig 10 Scallop dredge with spring-loaded tooth bar (Chapman *et al*, 1977)

the standard dredge in the Scottish (except Shetland) and Manx fisheries by the late 1970s. In 1980 a smaller dredge, only 2½ft (*ca* 0.8m) (*Plate 7*), was introduced into Scotland and the Isle of Man and has now become universally adopted except at Shetland, where the use of the fixed-tooth standard dredge persists.

Plate 7 Gang of 2½ft (0.8m) spring-loaded dredges

This smaller type, with teeth protruding up to 70mm, was first used in the recent English Channel fishery. In the eastern Channel, where the grounds are rough, they are used almost exclusively. They are towed from a towing bar, usually with a rubber bobbin at each end; the number fished ranges from three to six a side according to the size of boat. In the western Channel much of the fishing is by 'French-type' dredges (*Plate 8*). These are larger (up to 2m wide), heavier and have a diving plane which helps to keep them on the sea bed so that the teeth dig in better and make them more efficient on the smooth grounds on which they are used. Both the back and the belly consist of steel rings and the dredge is emptied by opening the rear end, which is normally kept closed by interlocking steel bars. Two to four dredges a side are fished from towing bars (Franklin *et al*, 1980).

When fishing for queens started in the Clyde Sea Area in late 1967, standard toothed dredges were used. At first local densities were so high that large quantities were caught by this gear. As densities became less, following the development of the fishery, it was essential to use more efficient gear able to catch queens which have swum up off the sea bed. In the Clyde this has taken the form of a modified otter trawl, fitted with a heavy ground rope with

Plate 8 French scallop dredge with diving plane *(By courtesy of A Franklin)*

rubber discs and bobbins and chaffing gear to protect the belly from the hard sea bed (see *Chapter 6*). This gear has remained the normal one for Clyde queen fishermen ever since. The fishing on Kish Bank off the east coast of Ireland also used otter trawls with the cod end protected by a sheet of scrap iron (Duff, 1976).

The short-lived Orkney queen fishery used beam trawls with a wooden beam, and the Shetland fishery uses a modification of the beam trawl which can best be described as a beam dredge, with a rigid steel frame and a netting bag (*Plate 9*).

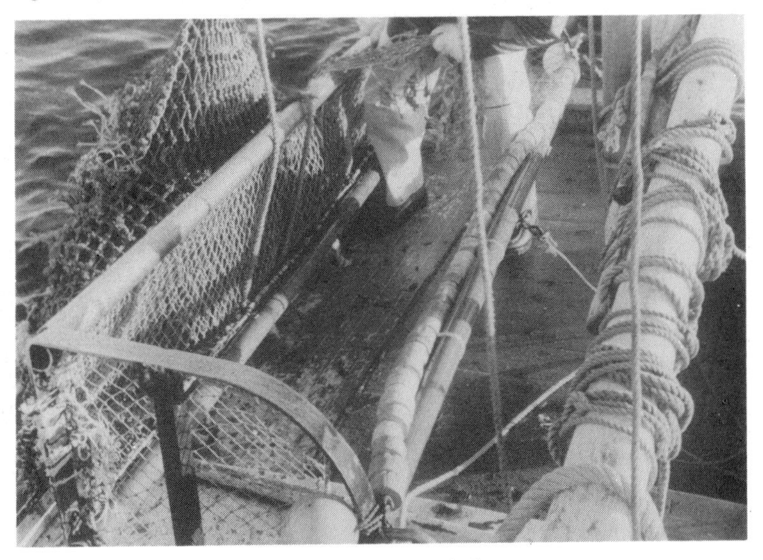

Plate 9 Shetland beam dredge

When queen fishing spread to the Irish Sea and the Isle of Man, where the rougher grounds were unsuitable for trawls, specially-designed queen dredges were designed by Manx blacksmiths. The Blake dredge (*Plate 10a*) was designed to catch queens as they swam up off the sea bed and leave dead shell and stones behind. It is large, up to 8ft (2.5m) wide, robust and heavy (180kg) and is consequently more difficult to handle. The frame is rather like that of a beam trawl, but made of heavy gauge steel, with two or three runners, and it has a heavy horizontal stone guard welded through the centre. It has a tickler chain, and instead of a blade it has a chain ground rope in front of a metal ring belly. The netting back of the bag starts the length of the runners ahead of the belly, and at the rear the back and belly are attached to a heavy, hollow metal tail bar. This dredge is used mainly by larger boats, towing one each side from a bridle, and is towed faster than the standard scallop dredge.

Plate 10a Blake queen dredge *(By courtesy of M S Rolfe)*

The Conolly roller dredge (*Plate 10b*) consists of a bag with the usual ring belly and mesh back, but has a toothless blade which is joined to a towing roller bar with a rubber wheel at each end. The mesh back extends right forward to the roller bar in order to prevent queens escaping. It is thought to be efficient on hard ground, but unsuitable on soft ground because the rollers sink in.

These designs of dredge were never adopted by the Scottish boats fishing for queens in the north Irish Sea. These boats have continued to use gangs of standard and spring-loaded toothed scallop dredges, though with shorter teeth, and with smaller belly rings (2¼in, 57mm, diameter) and back meshes owing to the smaller size of queens compared with scallops. The Manx boats have also now resorted to spring-loaded toothed dredges, which are easier to handle than the heavier queen dredges.

Plate 10b Conolly roller dredge (*By courtesy of M S Rolfe*)

Scallops are large, and very few below marketable size (90–100mm) are taken by the dredge. They are therefore usually sorted on deck by hand, and the dead shells, stones and other debris are thrown overboard. The queens are much smaller and the catch is much larger. Many undersized animals are taken and much more trash is caught owing to the smaller mesh and ring size used. Sorting by hand is therefore time-consuming, and various devices have been used to overcome this. Sloping tables have been arranged so that commercial sized queens can be picked out and rubbish allowed to roll overboard. Mechanical potato sorters have also been used, but they soon rust. However, during the past few years stainless steel rotary sorters have been developed and have proved satisfactory and their use has become widespread (*Plate 11*; Goodlad, 1976).

Plate 11 Rotary queen sorter (*Goodlad, 1976*)

A practice which developed in the late 1960s and has since expanded is the capture of scallops (though not the less valuable queens) by skin-divers. This lends itself particularly well to the west coast and islands of Scotland, where sheltered waters abound, and it is also carried on elswhere. The divers work in teams of from three or four upwards, using boats ranging from, say, as small as 16ft (5m) to fishing boats of 50ft (15m) or more.

The divers work on the sea bed in pairs, each diver carrying a net bag in which the scallops are placed. The bag might hold as many as 200 scallops and owing to its weight this is often supported by a float which can be made buoyant by the addition of more air.

In order to be profitable, divers' operations are limited to depths of 120ft (37m) or less, and an average of 90ft (27m) is more usual. In deeper water they would be unable to remain submerged for long enough to collect enough scallops. Operations are usually confined to inshore grounds, generally to areas where patches of sea bed suitable for scallops are interspersed with unsuitable rock and hard ground, areas in which dredging is not practicable, and so there is no conflict between diving and dredging. It is often said that those scallops which are inaccessible to dredging constitute a valuable breeding reserve and that heavy and indiscriminate exploitation by divers could endanger the stocks. However divers tend to gather only the largest scallops, over 120mm long, which give the best meat yield, and so still leave a residue which will have spawned several times before capture.

The proportion of Scottish scallop landings taken by divers has steadily increased, and in 1979 had reached almost 25%. The price fetched by diver-caught scallops is generally higher than of dredged scallops, exceeding £5.50 per dozen (£1,100 per tonne) in 1978, though it fell in 1979 to some £2.85 per dozen, this fall, like the general fall in prices, being due to the European market being flooded with cheap Japanese cultivated scallops.

The subject of commercial diving for scallops is dealt with in a recent book by Hardy (1981).

4
Handling, processing and marketing

After sorting on deck, the catch is hosed down to remove loose debris and packed in sacks each holding about 1cwt (50kg). Scallops are packed carefully, round valve downwards in order to retain as much water as possible, but the less valuable queens are shovelled in or fed in from the sieves. The catch is kept as cool as possible, often in the hold, especially if the boat is staying away from port for more than a day.

In the United Kingdom and Isle of Man fisheries, scallops and queens are usually consigned to a particular dealer or processor, scallops still being sold by the dozen (the price in 1980 was £1.50–3.00) and queens by the hundredweight. The consignee's lorry awaits the arrival of the boat or boats and the catch is despatched expeditiously. There is still a good market at Billingsgate for live scallops in the shell from the more accessible ports. Many landings are, however, made in remote places, and while some processing factories have been established in these remote areas (*eg* Shetland, Islay and Gairloch in Scotland), it is still necessary to transport some landings over considerable distances. For instance, scallops landed on the Moray Firth coast and queens landed at Weymouth on the south coast of England might be taken by lorry to be processed at Kircudbright in southwest Scotland.

The home market for scallops and queens is small, though growing. Most, however, are exported, either to the continent of Europe (mainly scallops) or America (mainly queens). In addition to scallops sold whole and fresh in the shell for the home market, some are sent whole, either fresh or blast frozen, to the continent of Europe, especially France and Spain, by refrigerated lorry and cross-channel ferry. Most scallops are exported as frozen meat, mainly with, but sometimes without, roe according to condition, and some are marketed as canned or breaded meats. Queens are mostly exported to the United States as frozen muscles.

The establishment of processing factories, first in the 1960s, revolutionized the scallop fishery and enabled the queen fishery to develop and expand. For the first time fishing became possible throughout the year, even in the warmest months. Scallops are best handled fresh on arrival at the factory, though in times of abundance they are deep frozen and stored for later use after thawing. Meat extraction, known as shucking, is a manual operation in the scallop and largest queens (*Plate 12*). The shell is opened by inserting a sharp knife along the inside of the flat valve. The viscera are then removed from the round valve, leaving the adductor muscle and roe attached and easily cut away. The whole operation, usually performed by teams of women, is carried out with amazing speed.

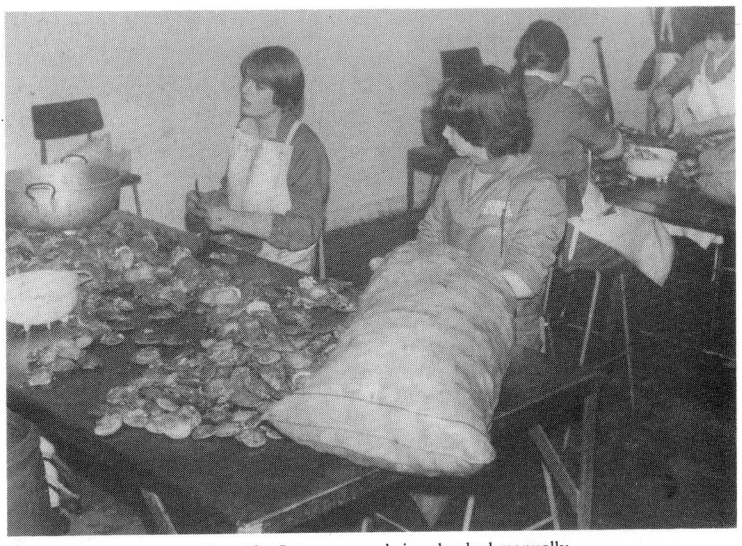

Plate 12 Large queens being shucked manually

While the live, shell-on market generally required large scallops, preferring those 5in (127mm) or longer, processing has provided an outlet for smaller specimens and is particularly beneficial in those areas, such as southwest England, where growth rates are low.

Because of their smaller size a quicker method of shucking

queens is desirable. The queens are scalded in boiling water for 40–50 seconds to open the shells and cooled by plunging into cold water. The flesh is then detached by shaking or by pushing with the finger, and the adductor muscle is pushed out from the rest of the viscera. Because the American market does not want the roe this has until recently been discarded. However, a roe-on market has recently developed for queens in Europe and so those with full roes are left with the roe attached to the muscle. Several processing factories are now equipped for the mechanical shucking of queens (*Plate 13*). The process involves scalding and shaking on a riddle to remove the meat plus viscera from the shell. The muscles are separated from the rest by means of a series of counter-rotating rollers. The gonad is discarded with the waste, and so mechanical shucking is unsuited to the preparation of a roe-on product, which is still done manually.

The meats of scallops or queens are washed, or even soaked in iced water if discoloured (Early and Stroud, 1981). They are then individually frozen, usually in single layers in trays in a blast freezer at –30°C. They are then best glazed by dipping in clean fresh water and packed, either in bulk or in catering or family-size packs. They keep in good condition for up to six months if stored at –30°C (Hardy and Smith, 1970).

The yield of scallop meat (muscle plus roe) varies according to the season from 10 to 16% of the whole weight in shell. The yield of queen meat (muscle only) ranges from 12 to 15%.

The price at first sale varies according to season. The weight of the adductor muscle and its protein and carbohydrate content are lowest in spring and build up to a maximum in the autumn (See *Chapter 8*). The scallop, however, fetches the best price when its gonad is fullest, in the spring in the English Channel, and in the spring and late summer/early autumn in the Isle of Man, Ireland and the west of Scotland.

Though the queen, and especially the scallop, in common with other shellfish, are luxury products, they are highly nutritious, being rich in glycogen and protein. Mason (1959b) quotes energy values of 87.1cal per 100g of muscle and 88.1cal per 100g of roe for scallops, which compare favourably with other shellfish.

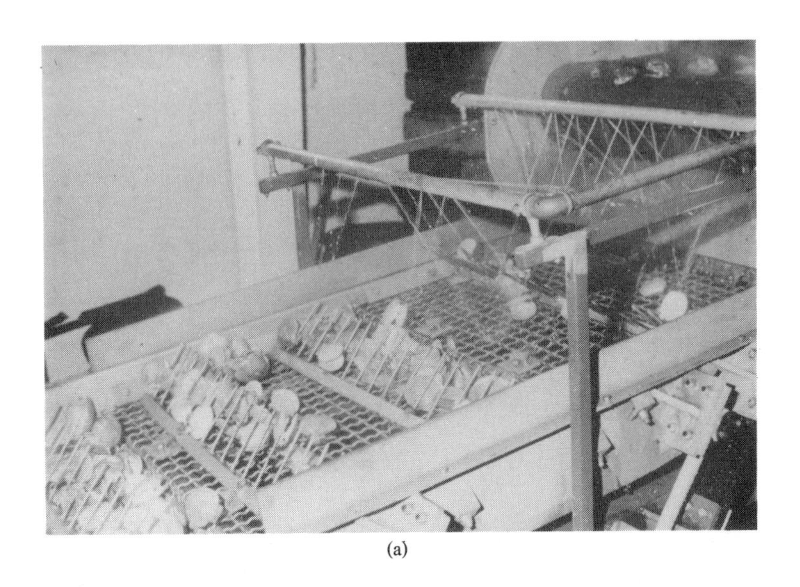

(a)

(b)

Plate 13 Mechanical shucking of queens: (a) Shaking to remove shell contents after scalding; (b) Separating the muscles by counter-rotating rollers (*By courtesy of A R Brand*)

5
Behaviour

While the scallop, *Pecten maximus,* and the queen, *Chlamys opercularis,* live on similar substrata and are often found together, their way of life and behaviour are quite different. Both species, in common with many other pectinids, are capable of swimming but the extent to which they do so differs markedly. Movement is effected by expulsion of water from the mantle cavity by rapid contraction of the adductor muscle and closure of the shell. The muscle consists of two distinct parts, closely bound together, a larger anterior part which can contract rapidly and is responsible for sudden movements of the shell, and a smaller posterior part which acts more slowly but more strongly and whose contraction enables the shell to stay closed against the effect of the elastic ligament. The type and direction of movement are dictated by the muscular edge of the mantle forming itself into one or more nozzles through which the water is ejected.

Pecten maximus normally lies recessed (*Plate 14, Fig 1.*) in a slight hollow in the sea bed, with its upper valve approximately level with or even just below the surrounding sea bed and covered with a thin layer of silt which makes observation by divers difficult (Baird and Gibson, 1956; Baird, 1958). It is usually detected by its habit of lying with the shell valves gaping so that the mantle curtain, eyes and tentacles are visible. Scallops which are recessed normally remain so for considerable periods, swimming only if disturbed. Those which have got onto a hard substratum unsuitable for recessing seldom remain in one position for more than a few days. They probably move at random and recess when they find a suitable substratum (Baird, 1958; Hartnoll, 1967).

In Strangford Lough, Ireland, Hartnoll found that *Pecten maximus* tends to orientate itself so that the inhalant current (in at the anterior and ventral margins) is helped by the prevailing uni-

Plate 14 Scallop recessed into the sea bed. The mantle curtain and tentacles are clearly visible (*Chapman et al, 1977*)

directional current. In more normal conditions with tidal reversal, Mathers (1976) found scallops orientated in such a way that the ventral part of the mantle curtain faced into either the flood or ebb tide and they did not change position with the change of tidal direction.

The process of recessing was described by Baird (1958). The scallop, lying on its right valve on the sand, repeatedly ejects a jet of water out of the right side adjacent to the auricle. This sometimes causes the scallop to swing round, but more commonly results in its becoming inclined to the sea bed. Once inclined, the scallop continues to squirt water by opening the valves and then closing them rapidly, the velum forming a barrier to the escape of water except in the direction desired. This blows a recess in the sand, and when it is large enough an exceptionally powerful squirt lifts the scallop round and up and it lands precisely in the recess. After a few days recessed the upper shell valve has acquired a coating of sand, often up to 5mm thick.

Baird considered that recessing is not a means of avoiding detection, as the tentacles and mantle are easily seen by the human eye at depths down to 37m, and the act of closing attracts attention. He considered that it might be connected with feeding, since recessing brings the inhalant current into or near the plane of the sea bed, thus facilitating the intake of detritus, which may be important as food.

Baird observed that the scallop's normal reaction to the approach of an object, such as a diver, was not to swim but to retract the tentacles and close the valves. Buddenbrock and Möller-Racke (1953) found, however, that the reaction of both scallops and queens to moving objects depends on their velocity, slow movement resulting in stretching of the tentacles towards the object and more rapid movement in retraction of the tentacles and closure of the valves. Little escape reaction was seen, and then mostly in young (0– and 1–group) animals. Apart from its use in finding a suitable substratum, swimming was also observed in the presence of starfish.

Thomas and Gruffydd (1971) classified the movements of *Pecten maximus* into non-escape movements and escape responses. The valves are normally open, with slight adjustments to maintain a steady flow of water through the gills for ventilation and food collection, though at intervals of 15–20 minutes sudden valve closures occur connected with expulsion of pseudofaeces. Step-like movements occur prior to large adjustments in the gape. More rapid valve closures occur in disturbed scallops. Escape responses can be induced by touching the scallop with an inanimate object, the human hand or a starfish, or by introducing starfish extracts near the edge of the mantle. The reactions result in either a movement of the scallop away from the stimulus or a partial or complete closure of the valves, presumably to protect the viscera. Response to starfish was produced experimentally only when contact was made with the extended tentacles.

Reaction was of three types, swimming (the common method), jumping and shell closure. Each swimming reaction was preceded by a wide gaping of the valves, and consisted of rapid clapping of the valves and expulsion of water on either side of the dorsal hinge so that the scallop moved ventral edge foremost, appearing to bite its way through the water (*Fig 11a*). Jumping featured a gradual relaxation of the adductor muscle followed by a more rapid relax-

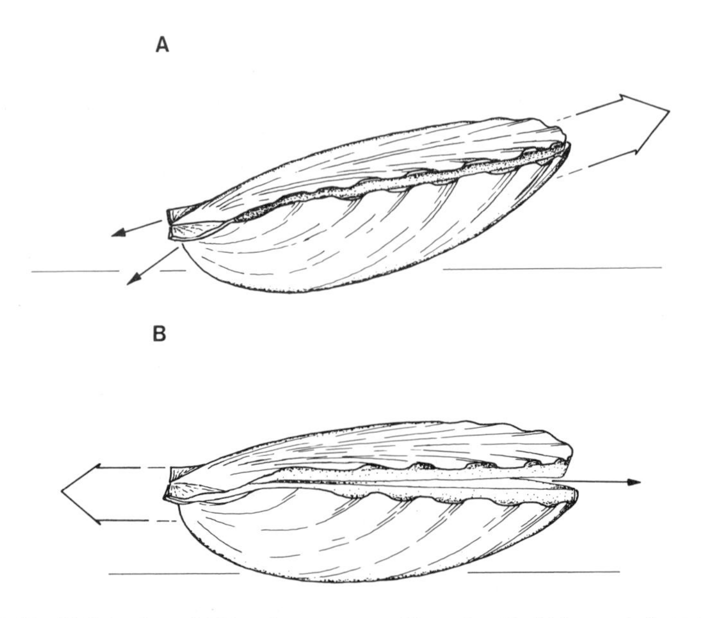

Fig 11 (A) Swimming and (B) jumping movements of the scallop. The thick arrow indicates the direction of movement and the thin arrows the water jets

ation and opening of the valves and even more rapid closure of the valves which expelled water ventrally and caused the scallop to jump hinge forward (*Fig 11b*), often being repeated several times in rapid succession. The scallop always settled between successive jumps. The simplest reaction was a partial or complete closure of the shell valves, followed by re-opening within a few seconds. The response protects the viscera and also serves to dislodge any starfish which is loosely attached to the shell.

Of seven starfish species used by Thomas and Gruffydd in their experiments only those predatory on molluscs, namely *Asterias rubens, Astropecten irregularis* and *Marthasterias glacialis* induced swimming and that not often. *Luidia ciliaris* and *Solaster papposus,* which are predatory but not on molluscs, produced jumping or closing reactions. The non-predatory species *Porania pulvillus* and *Henricia sanguinolenta* sometimes produced jumping or closure and sometimes no reaction at all, whereas contact with a glass rod always induced closure. This is seen as an adaptive reaction, escaping from predators but remaining in the depression when faced by non predators. Soemodihardjo (1974) reached similar conclusions with respect to the reactions of *Chlamys opercularis* to predatory starfish.

Thus swimming in pectinids is an escape response from potential predators. It is mostly induced by touch or chemically, though it can occur in response to the approach of fishing gear or divers up to 1–1.5m away. This is probably mediated by the eyes. Land (1966) showed that *Pecten maximus* responds to dark objects subtending 2° or more at the eye and moving through more than 2°. The swimming of one specimen of *C opercularis* often appears to induce swimming in others. If it is assumed that *Pecten* and *Chlamys* have similar powers of resolution, they should be able to detect the movement of others up to 1.5m away (Chapman, Main, Howell and Sangster, 1979).

The queen swims more actively than the scallop. This is attributed by Soemodihardjo (1974) to the greater ratio of cross sectional area of the quick adductor muscle to shell area in *C opercularis*. In both species young animals swim more actively than older ones. Swimming ability is hampered or even prevented by incrustations on the shell such as barnacles, serpulid worms, sponges and even fronds of seaweed, *Laminaria* (Rees, 1957; Chapman *et al,* 1979). The swimming endurance of *Chlamys opercularis* has been studied by Chapman *et al* (1979). Animals were stimulated to swim until fatigued by repeated tactile stimuli. Each stimulus at first produced a burst of swimming, but the distance swum and the duration of swimming declined with successive bursts, and most animals failed to respond at all after five bursts. The total distance travelled before exhaustion averaged up to 6.6m and the speed of swimming averaged 0.36 to 0.4m s^{-1}.

Fishermen have claimed that large scale migrations of queens occur, but, in view of their limited speed and endurance, directed migrations seem unlikely. Any large scale movement would be most likely to result from water currents acting on queens which had swum up off the sea bed. Chapman *et al* (1979) showed that in the first burst of swimming queens could reach a height of 0.31m, declining to 0.22m by the third burst. Minchin (1978a) found that young *P maximus* with one or two annual growth rings commonly achieved a height of 30–50cm and those of 120mm seldom exceeded a height of 10cm. Gruffydd (1976) showed that the shell shape of *C opercularis* generated enough lift to achieve neutral buoyancy at water current speeds of 0.3–0.4 m s^{-1}

(0.6–0.8 knots) so that dispersal by currents would be possible once the animal had left the sea bed (Chapman *et al*, 1979). At the end of each burst of swimming animals of both species fall passively to the sea bed with a to and fro motion (*Fig 12*). They normally land the correct way up, and Minchin (1978a) observed that small individuals of *Pecten maximus* fell more slowly than larger ones. If it should land the wrong way up or be inverted by any cause, the animal can right itself by somersaulting. It does so by suitably adjusting the mantle curtains and suddenly expelling water downwards at the free margins of the valves.

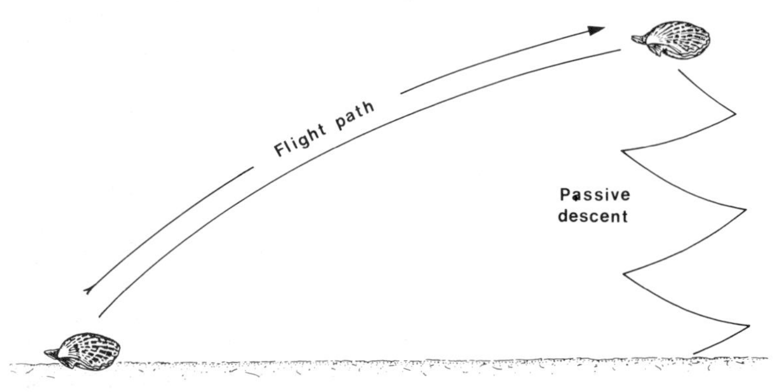

Fig 12 Swimming of the queen

Chapman *et al* (1979) found that *Chlamys opercularis* showed two tendencies in its reaction to a tactile stimulus; (*a*) to swim in the direction that the animal faces when resting (*ie* ventrally) and (*b*) to swim away from the stimulus. When touched, about half the animals swam ventrally in the direction they were facing and the other half pivoted round and swam away from the stimulus in the normal way. None moved hinge first in the manner described by Moore and Trueman (1971) in *C opercularis* and Thomas and Gruffydd (1971) in *P maximus* in response to predatory starfish.

6
Behaviour in relation to fishing gear and gear efficiency

Comparisons of the efficiency of various types of gear in catching scallops and queens were made by Mason (1970). The gears compared were: (*1*) a standard 4ft (1.22m) toothed commercial dredge with teeth protruding 6.4cm and spaced 7.6cm apart, belly rings of 83mm internal diameter and 75mm mesh netting back; (*2*) an Orkney beam trawl 8ft (2.44m) wide and 46cm high, with a tickler chain and a bag of 60mm mesh netting, and (*3*) a Manx beam dredge, 8ft wide and 28cm high, with a belly of 64mm rings and a back of 45mm mesh netting. All the gears were fished singly and therefore without a towing bar. The results (*Tables 1a, b*) showed that the toothed dredge was approximately five times as effective per unit width as the beam trawl in catching scallops and the beam trawl was approximately five times as effective as the toothed dredge in catching queens. Furthermore, the beam trawl was four times as effective as the Manx beam dredge in catching queens. Observations from a submersible and by aqualung divers showed that the variation was due to differences in the reactions

Table 1

(a) Catches of queens and scallops by scallop dredge and beam trawl

Gear	*Fishing time (min)*	*No. of queens caught*	*No. of queens per hour per foot width*	*No. of scallops caught*	*No. of scallops per hour per foot width*
4ft scallop dredge	891	1,729	29.1	152	2.7
8ft beam trawl	802	16,928	152.4	52	0.5

(b) Catches of queens by beam trawl and Manx queen dredge

Gear	*Fishing time (min)*	*No. of queens caught*	*No. of queens per hour per foot width*
8ft beam trawl	120	4,508	281.8
8ft Manx queen dredge	105	986	70.4

of the scallop and queen to the approaching gear. Most scallops on the one hand tended to close the shell and sink into their recesses, so that, while the teeth of the standard dredge could dig them out, the other two gears tended to pass over them. The queen on the other hand reacted by swimming off the sea bed when the gear was still 30–60cm away and observations showed dense clouds of them passing over and round the gear. The standard dredge, having a lower (23cm) mouth opening than either the beam trawl (46cm) or the Manx beam dredge (28cm), would thus catch fewer of the swimming queens.

Despite its relative inefficiency, the toothed dredge, usually now spring-loaded, is used in the queen fishing prosecuted by Scottish boats from the Solway coast, especially Kirkcudbright, in the Irish Sea. This is said to be because of the rougher nature of the grounds fished there, the beam trawl being easily damaged and the large Manx beam dredge rapidly filling with stones.

The most effective gear used for queen fishing, because of its great vertical mouth opening, is the otter trawl, a modified form of which is used in the commercial queen fishing in the Clyde Sea Area. However, the limited swimming ability of the queen has an important bearing on the effectiveness of the trawl gear (Chapman *et al,* 1979). The net studied was typical of those in commercial use, a light trawl (Cosalt 200hp) with 14.6m headrope and 19m fishing line. Standard 1.5 x 0.9cm metal Vee otterboards were joined to the wings of the net by short (2m) chain sweeps. The heavy groundrope had 150mm diameter rubber discs and large 300mm diameter rubber bobbins at the centre. The belly was protected by chaffing gear owing to the nature of the sea bed, which consisted of sand, stones and occasional boulders. The towing speed was estimated as 2 knots (1m s^{-1}) (*Fig 13*).

The animals were usually induced to swim when touched, but sometimes they responded earlier, this reaction probably being mediated by the eyes. They were caught when they swam up and passed over the fishing line of the approaching trawl. Only 30% of the queens in the path of the gear were caught by the standard trawl, the remainder being overrun by the gear. Many were first disturbed by the otterboards and wings, most reacting 0.3–1.3m ahead of the boards. Some escaped in the flow of water over the board while others swam into the path of the net. Few swam high enough to escape over the headline (1.5m high at the centre). The

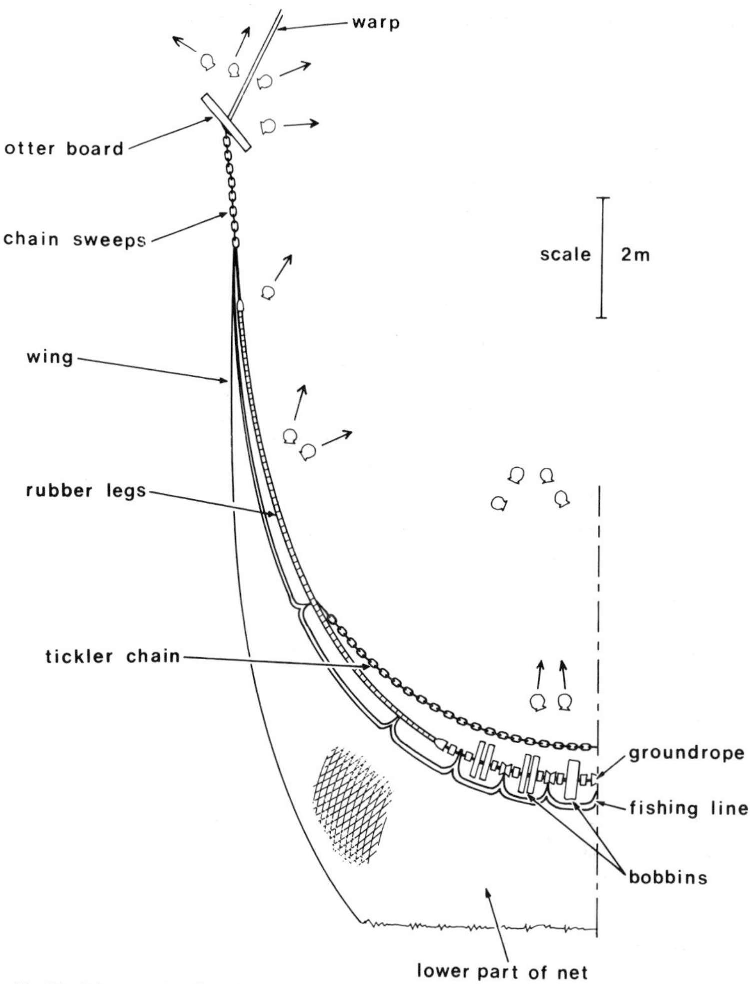

Fig 13 Diagram showing the dimensions of a queen trawl and some observed orientations of queens to various parts of the gear. (*Based on Chapman et al, 1979*)

animals swam away from the boards for a short distance (1.2m) and then resettled until approached by other parts of the trawl. By the time the fishing line reached them most animals had undergone several short bursts of swimming. Many of these failed to respond to the groundrope and passed under the fishing line. Modifying the trawl by fitting a tickler chain induced some animals to swim in front of the fishing line and increased the percentage of animals caught to 64.

Animals lying near the centre of the net's path, and therefore not previously stimulated by the boards, reacted in advance of the groundrope with ample time to swim up into the net. Others, reacting first to the otterboards and wings, were stimulated to swim several times before arriving at the centre of the fishing line and were by then incapable of further swimming and so were not caught, while yet others reacted only when touched by the gear. Some of these latter were caught after being touched by the tickler chain which was added as a modification. A tickler is generally fitted to beam trawls and dredges used for catching *C opercularis* (Mason, 1970; Rolfe, 1973) but may not be practicable on the rough grounds being fished by otter trawl. It should, however, be possible to achieve a similar improvement in catch by fitting the groundrope further in front of the fishing line.

The fate of individual queens stimulated by the forward parts of the gear depends partly on the direction in which they move relative to the path of the trawl. The greater the tendency to swim in the same direction as the net, the more bursts of swimming will be accomplished before the groundrope reaches them and the lower the probability of capture. Queens tend to orientate themselves with the shell gape facing into the direction of the tide and to swim in the direction they are facing. Thus a high proportion may swim towards a net towed with the tide and this should result in larger catches. The importance of vision in reaction to fishing gear is indicated by the fact that catches by day generally exceed those at night. It is best to tow the net sufficiently fast to keep the warp and bridles off the sea bed in order to avoid disturbing the queens too far ahead of the net.

Despite being commonly used in fisheries for *Pecten maximus,* the standard toothed dredge is a relatively inefficient piece of gear. Baird (1955, 1959) estimated that only between 5% and 20% of the scallops in the dredge's path are caught. Baird found that owing to tightening and slackening of the warp, the standard dredge proceeds in a series of shallow leaps interspersed with short spells of effective fishing. Baird designed a dredge to overcome this difficulty, which glides along the sea bed on runners, has a toothed bar set at a smaller angle (45°) to the sea bed and teeth which project a shorter distance, so that they scrape up the scallops without digging as far into the sea bed. Further, it has a hydrofoil plate which tends to depress the dredge onto the sea bed

(*Plate 15*). It was claimed that it can be towed faster than the standard dredge and that it is two to three times as efficient on clean sandy ground. Rolfe (1969), using two different methods, obtained estimates of 24% and 33% for the efficiency of the Baird dredge in catching scallops $\geqslant$89mm long. It has not, however, given the same increase in efficiency on a soft gravel bottom. On rough ground, moreover, the Baird dredge tends quickly to fill up with stones and so has not been adopted by commercial fishermen in the British Isles.

Recent attempts to obtain more accurate estimates of the effici-

Plate 15 Baird runner dredge (*By courtesy of M S Rolfe*)

ency of the standard (and spring-loaded) toothed dredge have been reported by Chapman, Mason and Kinnear (1977) and Mason, Chapman and Kinnear (1979). This work was done by divers examining directly the path of the dredge on a smooth sandy bottom and comparing what was caught with what was left behind. Most scallops left behind were dug out and only 16% were still recessed.

Both the tooth spacing (Baird and Gibson, 1956) and the ring and mesh size (Drinkwater, 1974) are selective agents in the toothed dredge. In the recent experiments the scallops were divided into those $<$ 80mm and those $\geqslant$ 80mm long, a size which corresponds approximately to the tooth spacing and mesh and ring size. Only 3.3% of scallops $<$ 80mm long were caught and in two separate sets of experiments means of 27.0% and 23.6% of scallops $\geqslant$ 80 mm long were caught (*Fig 14*).

The teeth of the standard dredges dug about $\frac{1}{2}$–$\frac{3}{4}$ of their length into the sea bed and generated a mound of sediment in front of the toothed bar, most of which was deposited round the edge of the dredge. At times the mound completely filled the dredge opening (*Plate 16*), particularly when large stones or shells blocked some of the spaces between the teeth. Most of the scallops were pushed aside, though a few small ones ($<$ 90mm)

Plate 16 Mound of sediment generated by the toothed dredge (*Chapman et al, 1977*)

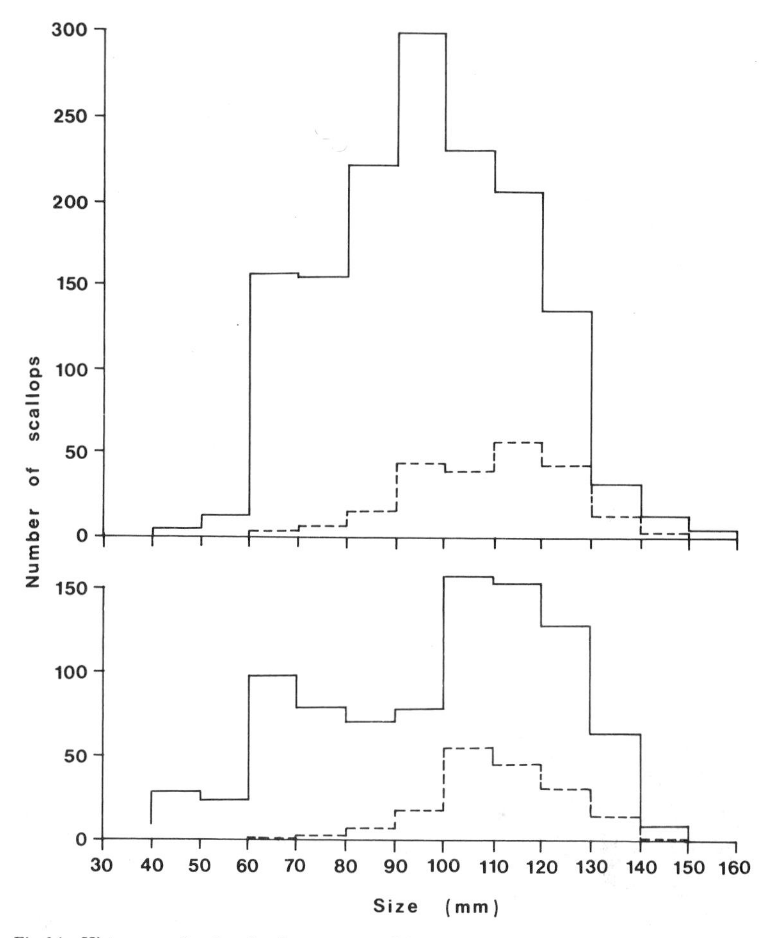

Fig 14 Histograms showing the size structure of the scallop population (————) and the proportion of each size caught by the dredge (– – – – –) in two experiments in the Clyde Sea Area. (*Chapman et al, 1977*)

disturbed by the gear were seen to swim away. In an attempt to retain the scallops pushed aside, a dredge was modified by the addition of metal lattice wings. However these, while they reduced the amount of sediment pushed aside, created an even greater mound of material in the mouth of the dredge, resulting in 30% more trash being taken without increasing the catch of scallops (Chapman *et al,* 1977).

Two other modifications have been tested (Strange, 1979). One dredge had a trash gap between tooth bar and bag in an attempt to allow trash to escape. A second was fitted with a triangular tooth bar in an attempt to allow the mound of trash to be pushed away or escape through the triangular gap behind the bar. In tests against a standard dredge the catch of both the modified dredges was inferior. It appears that scallops as well as trash escaped.

The spring-loaded dredges with thinner and longer teeth which penetrated less far into the sea bed took less trash but were less efficient than the fixed bar dredge, taking 15% of scallops $\geqslant$ 80mm and only 2% of those $<$ 80mm. It should be emphasized that these results were obtained on a smooth sandy bottom – the results might well be different on the rough bottoms for which the spring-loaded dredge was developed.

The efficiency of the standard dredge varied with the size of

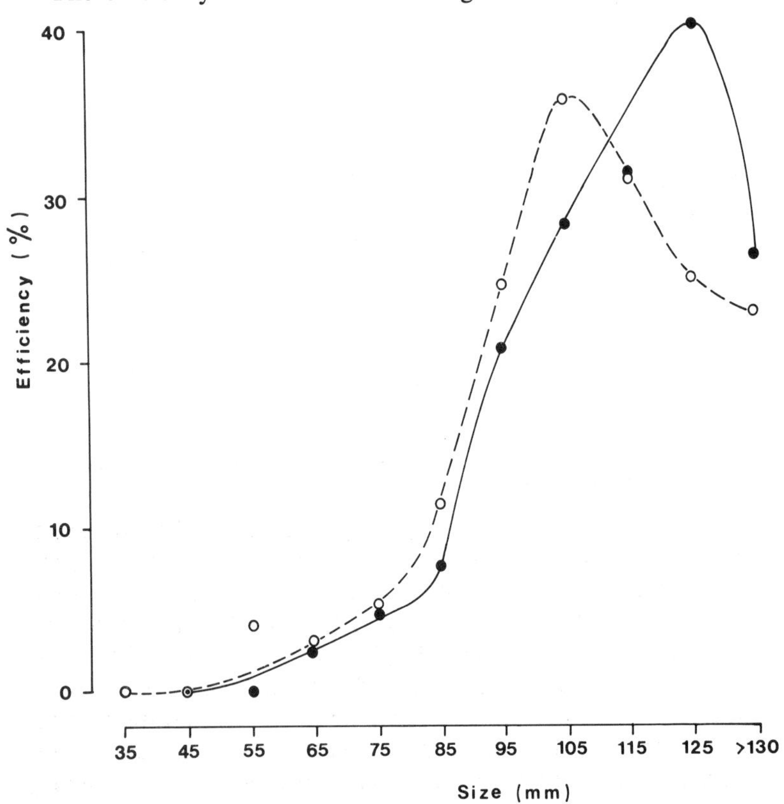

Fig 15 Efficiency of the scallop dredge for different sizes of scallop as shown by the percentage caught at each size group in two different experiments. (*Chapman et al, 1977*)

scallops caught, rising from zero at 45mm length to a peak of 35% at 105mm in one set of experiments and 40% at 125mm in another, and then declining with further increase in size (*Fig 15*). It is likely that the overall efficiency in relation to size of scallop is a function of (*a*) selectivity, S, of the teeth and meshes, which influence the retention of the smallest scallops, and (*b*) the efficiency of capture, E, which becomes increasingly operative for larger scallops which are hindered by the mound of sediment from entering the dredge (*Fig 16*).

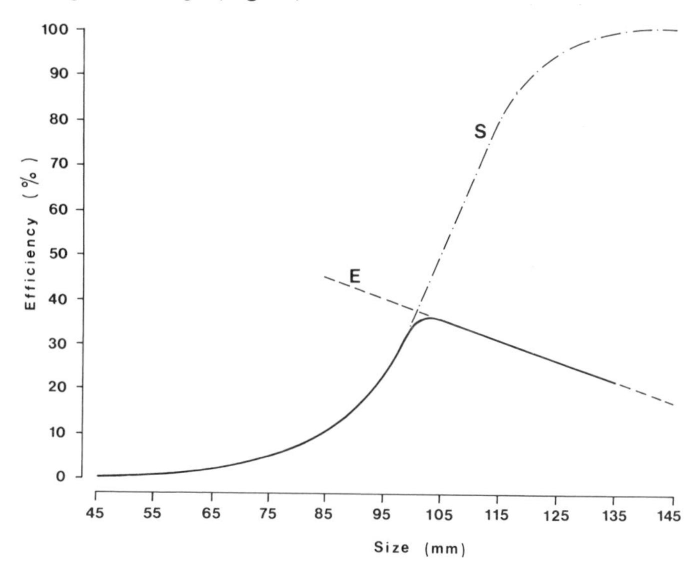

Fig 16 Diagram to show the efficiency of the standard scallop dredge (solid line) made up of two components, selectivity (S) and catching efficiency (E). (*Chapman et al, 1977*)

While one would expect the youngest scallops to be most abundant, a scarcity of young, and therefore small, scallops in commercial dredge catches has been reportd by a number of workers (Baird, 1952; Baird and Gibson, 1956; Mason, 1957). Recent work has tended to discount the idea that the younger (0+ to 2+) animals settle and live on feeder beds and later migrate onto adult beds. It is most likely that the scarcity is due to selectivity of the dredge, since divers have caught many more smalls than the dredge and even the divers must miss some owing to the difficulty of observing the smallest. After years of poor settlement (Mason, 1957) 0+, 1+ and 2+ animals would be even scarcer than usual.

7
Breeding

The family Pectinidae is interesting in that it contains both unisexual and hermaphrodite species. Both the scallop and the queen are hermaphrodite.

While the process of breeding is basically similar in *Pecten maximus* and *Chlamys opercularis*, differences do occur which necessitate separate descriptions for the two species. Even within one or other species differences occur from place to place and from time to time. I shall therefore first describe the breeding of *P maximus* in Manx waters as shown by my own detailed study (Mason, 1958a) and then show how the basic pattern varies.

The breeding of *Pecten maximus*

The study was based on an examination, both macroscopic and microscopic (from stained sections and smears), of samples of scallops of all ages up to 13 years taken at roughly weekly or fortnightly intervals from a depth of 24–29m (13–16fm) off Port Erin, Isle of Man, during the period October 1950–October 1952.

The structure of the gonad

The single gonad is posterior and ventral to the rudimentary foot, forming a tongue-shaped mass attached to the adductor muscle. The proximal part is white and forms the testis; the ovary is orange-red and lies distal to the testis. A loop of the alimentary canal passes through the gonad, penetrating into the ovary. This loop cannot be seen in the mature gonad unless, as occasionally happens, it passes close to the free, ventral edge of the ovary.

The boundary between testis and ovary is usually quite sharp, though irregular, but sometimes islets of tissue of one kind occur within the tissue of the other kind. This irregularity in the distribution of the spermatic and ovarian tissue occasionally reaches a state in which the gonad is almost exclusively either male or female. I found two gonads composed entirely of female tissue.

The gonad is made up of many branched, ciliated tubules or ducts, which bear sacs, the alveoli or follicles (*Plates 17, 18*). The ducts are round in cross-section, 70μm or more in diameter. The lumen may be as small as 20–25μm, and is lined with ciliated epithelium. In serial sections the ducts may be traced and are seen to join and rejoin. They eventually join one of the two main ducts, one on each side, which are much wider, up to 1,000μm in section, and open dorsally, one into each kidney.

The sexual products arise by proliferation of the germinal cells which line the follicle walls. The follicle wall is usually less than 1μm thick. The follicles become filled with sexual products, and those near the surface of the gonad appear to the naked eye as small, rounded, red or white bodies; the full follicles range in size from 300 to 700μm.

In each follicle all the contents are normally of one sex, though in one gonad follicles were found containing both male and female elements, spermatozoa, spermatocytes and large oocytes being found side by side (*Plate 17*). The male follicle usually contains a few early stages of spermatogenesis near the follicle wall, but the lumen is filled with spermatozoa. Each spermatozoon has a minute conical head, about 1.4μm, which stains intensely in

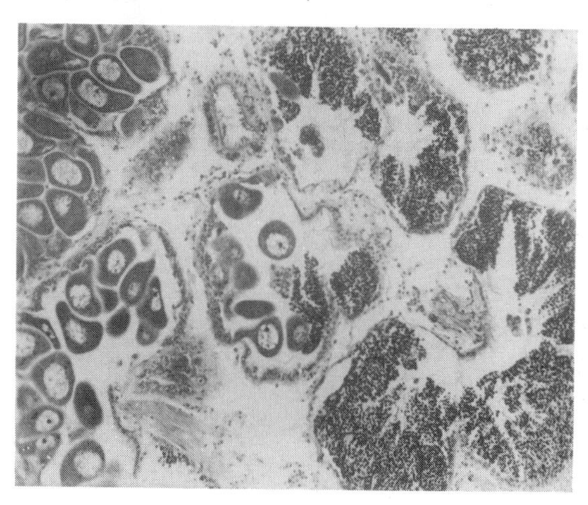

Plate 17 Transverse section of abnormal gonad of *Pecten maximus*, showing ambisexual follicles (magnified x 94) (*Mason, 1958a*)

haematoxylin, and a tail about 50μm long. The spermatozoa are arranged radially from the centre of the follicle, or from the point where the follicle opens into the ciliated duct, with their tails pointing towards the centre or towards the duct. The female follicle contains a few young oocytes attached to the wall, while the lumen is full of large oocytes. The large oocytes appear polygonal in sections, and are packed tightly in the follicles; their greatest diameter in sections is about 80–90μm. The large oval, or spherical, vesicular nucleus has a diameter roughly two-thirds that of the cell. A delicate network of chromatin fibres extends through the nucleus, and one conspicuous acentrally-placed nucleolus is present. The nucleus is surrounded by granular cytoplasm, which often contains bodies which stain purple in haematoxylin. The whole cell is surrounded by a membrane some 1.5μm thick.

Little connective tissue is present between the follicles of the full gonad although some is present round the loop of the alimentary canal and the ducts, and sometimes near the outer wall of the gonad. The connective tissue consists mainly of a network of fibres. Also between the follicles are transverse muscle fibres.

The outer wall of the gonad consists of two layers, an outer epithelial and inner muscular layer. The muscles of the wall, together with the transverse muscles, probably assist the ciliated lining of the ducts in ejecting the genital products.

The breeding cycle

In Manx waters, scallops first spawn in the autumn following the deposition of the second growth ring on the shell, when most of them are about two years old (Mason, 1957) (see *Chapter 8*). In the following year they have one main spawning, in autumn, and thereafter they have two main spawnings each year, one in spring and one in autumn. Scallops which have never spawned are called virgins, between the first and second spawnings they are called juveniles, and after the second spawning adults.

In order to determine the breeding cycle, gonads from the regular samples of scallops were classified into eight arbitrary stages of maturity, stages 0–VII, which are described in *Table 2*, the colour nomenclature being that of Ridgway (1912). The external features of the various gonad stages are shown in *Fig 17*, and their microscopic structure in *Plate 18*. Since the duration of the act of spawning was found to be so short as to make it most unlikely that many scallops in the spawning condition would be observed in nature, no spawning stage is included.

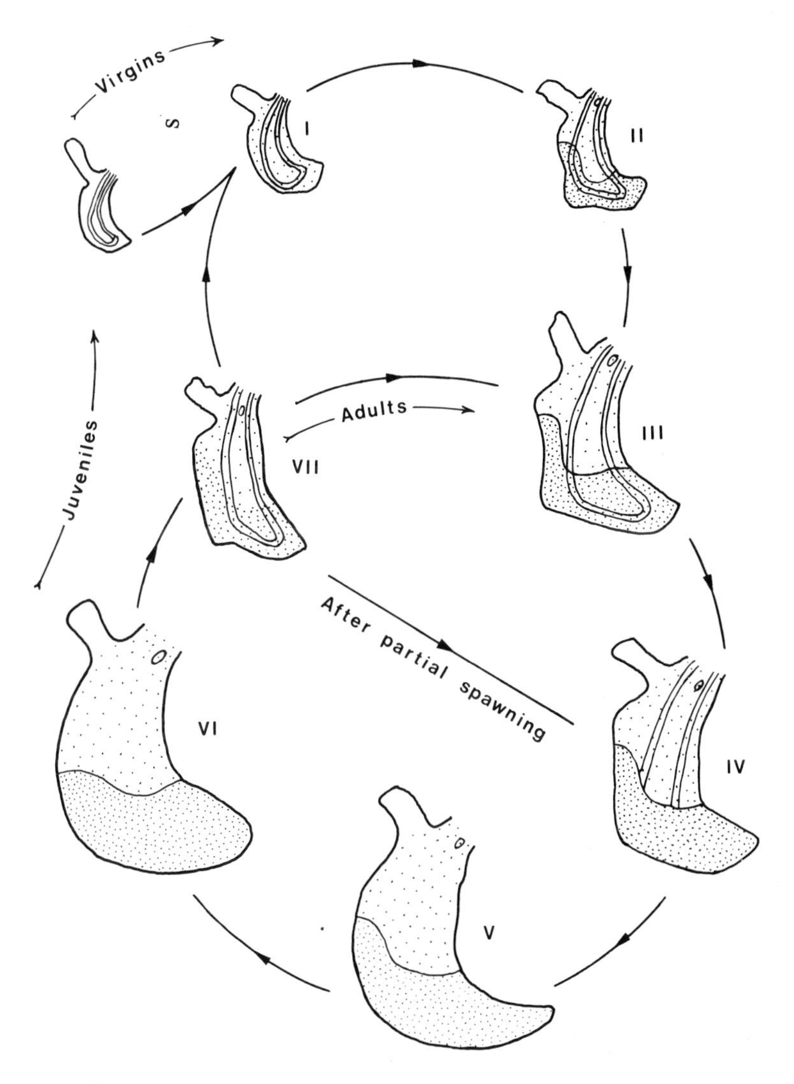

Fig 17 Macroscopic changes in the gonad of *Pecten maximus*. (*Mason, 1958a*)

Table 2 *Pecten maximus*. Macroscopic and microscopic changes in the gonad (Mason, 1958a)

Gonad stage	*External features (Fig 17)*	*Histological details*
0 Immature (virgin)	Gonad small, flat and angular, transparent and colourless. No reproductive tissue visible to naked eye. Whole of loop of alimentary canal clearly seen	Youngest scallops caught (2–3 months old) showed beginning of development, which then continues. Gonad at first almost completely occupied by loop of alimentary canal, but later connective tissue develops. Narrow tubules, bearing primary germ cells on walls, ramify and give rise to follicles and ciliated ducts, rounded in section. Primary germ cells irregular, 8–14μm, with oval, vesicular nucleus and scattered chromatin, lightly stained in haematoxylin. Follicular cells also present. Primary germ cells give rise to gonia, each with a spherical, vesicular nucleus, 3.5-7μm; chromatin scattered, but stains more densely than primary germ cell; well-defined nucleolus; usually a little cytoplasm. Formation of gametocytes, shown by synapsis, occurs first in distal (female) part of gonad (synapsis shows as clumping together of chromatin into an irregular, densely-staining mass). Gonad increases in size with formation of follicles and connective tissue (*Pl 18A, B*).
I Developing (virgin) or spent-recovering (juvenile)	Gonad growing, still flat and angular. Reproductive tissue now visible to naked eye as minute follicles, translucent and sparse. Gonad uniformly fawn-coloured, with no visible differentiation into testis and ovary. Alimentary canal visible	*Developing (virgin).* Follicles growing. Synapsis now occurs in proximal (male) part of gonad. Male follicles lined by several layers of spermatogonia, and lumina filled with synaptic and post-synaptic spermatocytes, with occasionally a few spermatids. Spermatocyte has a little cytoplasm, and appears polygonal in sections; nucleus roughly spherical 2.5–3.5μm, unevenly distributed chromatin, fairly densely stained, no visible nucleolus. Spermatid 2μm, spherical nucleus, chromatin evenly or somewhat unevenly distributed. Female follicles have oogonia and synaptic oocytes near walls, and young oocytes up to 30μm, growing rapidly, in lumina. Young oocytes have somewhat fibrillar cytoplasm; nucleus spherical, vesicular, with a delicate chromatin network and spherical nucleolus. Spaces in female follicles. Much conective tissue between follicles (*Pl 18C, D*). *Spent-recovering (juvenile)* is similar, but has larger spaces in follicles. Before recovery becomes obvious to naked eye many gonia are produced from residual gonia and primary germ cells, and synapsis occurs as above.
II Differentiated gonad (virgin and juvenile)	Gonad growing, still flat and angular, now obviously differentiated into testis and ovary; male follicles white and female fawn or light salmon orange, colour imparted by contents. Follicles still small and sparse, and alimentary canal still visible	Spermatozoa appear at centre of male follicles; many synaptic and post-synaptic spermatocytes and some spermatids; several layers of spermatogonia near walls. A few oogonia and synaptic oocytes on wall of female follicles; many half-grown (30–60μm) oocytes, appearing stalked, with granular cytoplasm, and a few young oocytes. Less connective tissue, but still continuous between follicles (*Pl 18E*).

III Recovering	Gonad larger and proportionately thicker, angular. Flabby, containing free water, especially in adults after spawning. Assuming brighter colour, testis white and ovary bittersweet orange. Follicles larger and denser, but still spaces between them, and alimentary canal still visible	Male follicles contain more spermatozoa, not yet closely packed, arranged radially; still many spermatogonia and spermatocytes near walls, and some spermatids. Few oogonia and synaptic oocytes in female follicles, and lumina almost filled with growing oocytes, mostly half-grown, a few larger and smaller. Few spaces left in follicles, rather more in adults (in which stage III is the first recognizable stage of recovery after spawning) than in virgins and juveniles. Connective tissue still present, but disappearing. Main genital ducts becoming flattened (*Pl 18F*).
IV Filling	Gonad still larger and thicker (thickness about $\frac{1}{3}$ width); still somewhat flabby, containing a little free water. Outline less angular, but not completely smooth. Colouring brighter due to denser colouring of follicles, testis white, ovary bittersweet orange or grenadine pink. Follicles larger and closer together, the latter especially in ovary. Alimentary canal still visible between follicles in testis, but not in ovary, but its outline still discernible owing to thinness of gonad	Spermatogonia still form a continuous layer on walls of male follicles, and inside them a band of spermatocytes and a few spermatids; lumina contain many spermatozoa arranged radially. Few oogonia and synaptic oocytes in female follicle; fewer young oocytes; lumen contains more half-grown and a few almost fully grown (60–80μm) oocytes. Little connective tissue except round alimentary canal and ducts. Main ducts larger and more flattened (*Pl 18G, H*).
V Half-full	Gonad again larger and thicker, firmer, and containing very little free water. Rounded, with tapering tip. Brighter, testis creamy-white, ovary grenadine pink or grenadine. Follicles larger, becoming packed together. Loop of alimentary canal visible, but still causes wall of gonad to bulge	A few spermatogonia and spermatocytes remain near walls of male follicles, lumina full of spermatozoa which are becoming closely packed. Walls of female follicle lined with a few young and half-grown oocytes; lumina filled with almost fully grown oocytes, each with germinal vesicle still intact; very few oogonia and synaptic oocytes, indicating that production of oocytes has now almost ceased. Very little connective tissue.
VI Full	Gonad is now at its largest, thickest (thickness about $\frac{1}{2}$ width) and firmest, containing no free water. Rounded to tip. Bright, with follicles highly coloured and closely packed; testis cream coloured, ovary usually grenadine. Loop of alimentary canal indiscernible	Follicles maximum size. Male follicles packed with spermatozoa, still arranged radially; few scattered spermatogonia and spermatocytes persist near walls. Female follicles packed with fully-grown (80–90μm) polygonal or pear-shaped oocytes, whose germinal vesicles show signs of breaking down; very few scattered oogonia and young oocytes persist near walls. No connective tissue except round alimentary canal and ducts. Main ducts flattened (*Pl 18J, K*).
VII Spent and partially spent	Spawning may be partial or complete. Gonad dull, angular, thin and collapsed; flabby, containing much free water. Spent gonad fawn-coloured and loses visible differentiation into testis and ovary after spawning for first time. Older scallops usually retain differentiation, testis yellowish-brown, ovary dull orange pink. Follicles appear empty. Partially spent gonad always retains differentiation; testis yellowish white, ovary dull, bittersweet orange or orange chrome. Follicles appear hollow, with a coloured ring round periphery indicating retention of some genital products	Follicles smaller and collapsed, containing large spaces. Main ducts wide. Some connective tissue visible. *Spent gonad* after spawning for first time contains a few residual primary germ cells, spermatogonia and spermatocytes on walls of male follicles, but few or no spermatozoa; female follicles have a few primary germ cells, oogonia and young oocytes on walls. Older scallops retain more spermatocytes and a few spermatozoa, and more young and some half-grown oocytes (*Pl 18L*). *Partially spent gonad* retains more genital products. Many residual spermatocytes and spermatozoa, and half-grown and almost fully grown oocytes (*Pl 18M*).

Plate 18 Photomicrographs of transverse sections through gonads of *Pecten maximus* in various stages of maturity (all magnified x 94) (*Mason, 1958a*)

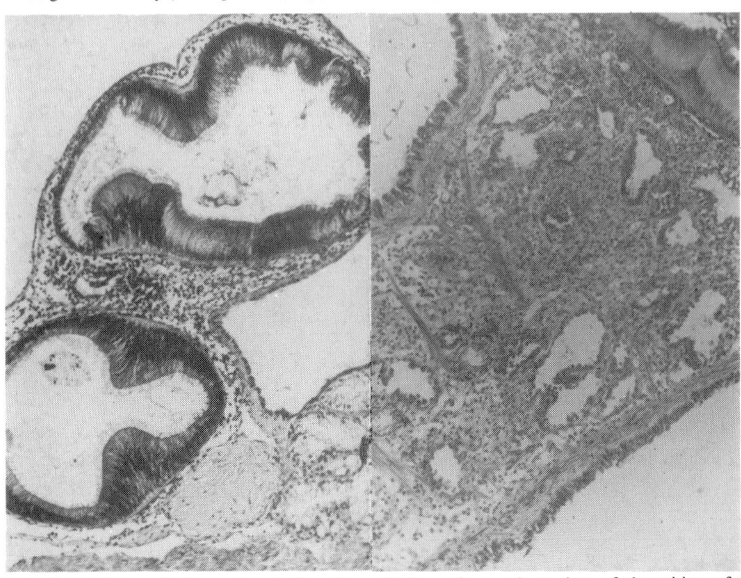

A Stage 0 gonad, December before first growth ring

B Stage 0 gonad, at time of deposition of second growth ring (April). Gonia showing as thickenings on edges of follicles

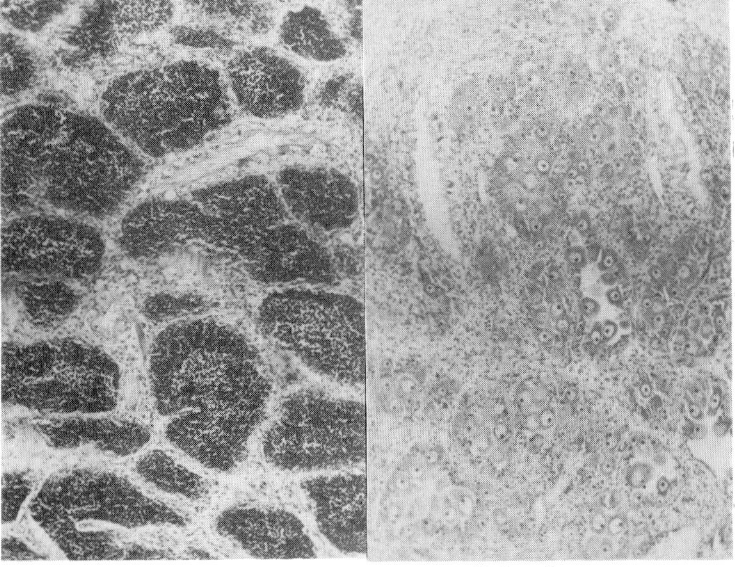

C Stage I testis

D Stage I ovary

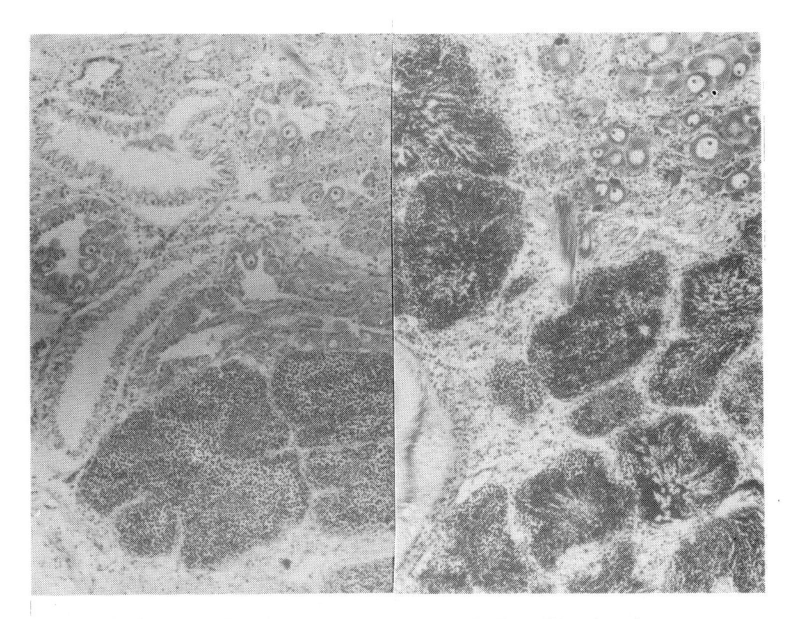

E Stage II testis and ovary

F Stage III testis and ovary

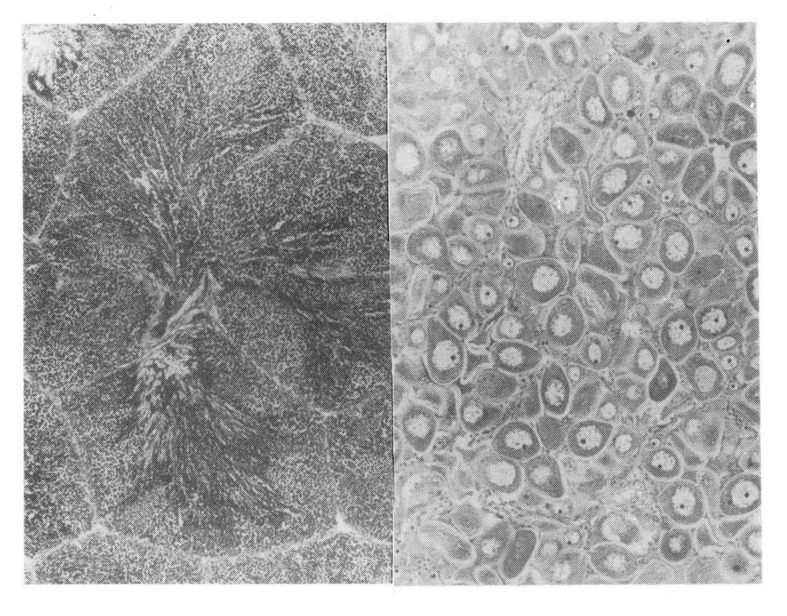

G Stage IV testis

H Stage IV ovary

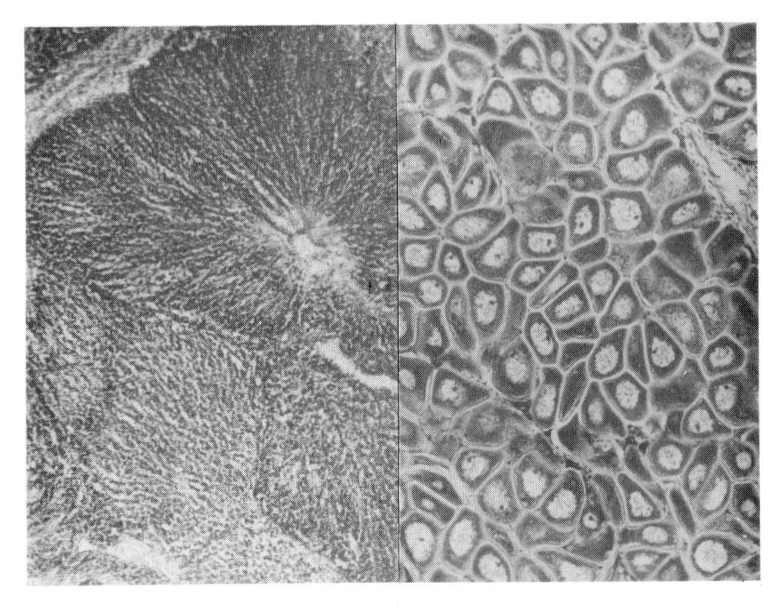

J Stage VI testis

K Stage VI ovary

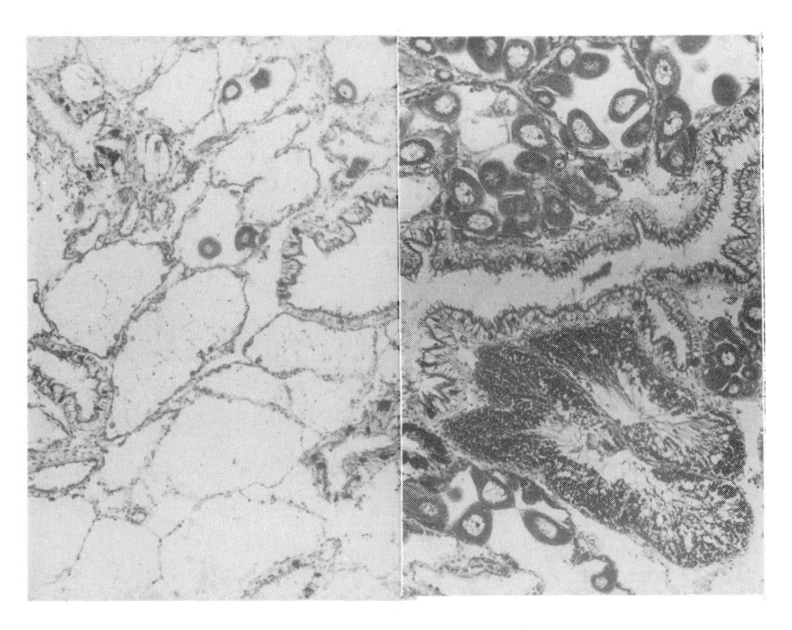

L Stage VII testis and ovary (partially spent)

M Stage VII testis and ovary (spent)

The virgin scallop *Table 3* shows the monthly percentages of gonads of virgin scallops in each stage of maturity during the first two years or so of the scallop's life.

Table 3 Monthly percentages of virgin scallops at each stage of maturity during the first 2-2½ years of life (Mason, 1958a)

(Owing to the small numbers caught, data for scallops caught in corresponding months of the period of the investigation (*eg* March, 1951 and March, 1952) are combined, and all the scallops are treated as if they were spawned in the same year)

	Gonad stages								
Month	*0*	*I*	*II*	*III*	*IV*	*V*	*VI*	*VII*	*No. of gonads*
Apr	–	–	–	–	–	–	–	–	–
May	–	–	–	–	–	–	–	–	–
June	–	–	–	–	–	–	–	–	–
July	–	–	–	–	–	–	–	–	–
Aug	–	–	–	–	–	–	–	–	–
Sep	–	–	–	–	–	–	–	–	–
Oct	100.0	–	–	–	–	–	–	–	2*
Nov	–	–	–	–	–	–	–	–	–
Dec	100.0	–	–	–	–	–	–	–	1
Jan	100.0	–	–	–	–	–	–	–	2
Feb	100.0	–	–	–	–	–	–	–	12
Mar	100.0	–	–	–	–	–	–	–	17
Apr	100.0	–	–	–	–	–	–	–	3†
May	100.0	–	–	–	–	–	–	–	9†
June	100.0	–	–	–	–	–	–	–	5
July	100.0	–	–	–	–	–	–	–	2
Aug	100.0	–	–	–	–	–	–	–	5
Sep	100.0	–	–	–	–	–	–	–	6
Oct	100.0	–	–	–	–	–	–	–	12
Nov	100.0	–	–	–	–	–	–	–	23
Dec	100.0	–	–	–	–	–	–	–	18
Jan	100.0	–	–	–	–	–	–	–	15
Feb	97.7	1.2	1.2	–	–	–	–	–	85
Mar	92.3	5.5	2.2	–	–	–	–	–	91
Apr	40.2	27.8	8.2	12.4	11.3	–	–	–	97‡
May	22.7	14.7	28.0	20.0	10.7	4.0	–	–	75‡
June	10.2	28.6	16.3	10.2	16.3	6.1	12.2	–	49
July	–	17.4	21.7	17.4	30.4	8.7	4.3	–	123
Aug	–	8.3	13.9	8.3	8.3	5.6	5.6	50.0	36
Sep	–	–	–	8.1	13.5	5.4	9.5	63.5	37

* Millport scallops † 1st growth ring ‡ 2nd growth ring

All scallops with no growth rings and almost all with one ring possessed stage 0 (immature) gonads, and the first sign of development obvious to the naked eye appeared in the late winter or early spring about the time of deposition of the second growth ring, although development of tubules and germ cells had been going on continuously since a very early age (*Table 2*). Development continued throughout the spring and summer, through stages I–VI, and culminated in the act of spawning in the August or September after the deposition of the second growth ring. Spawning was complete in most individuals, and involved the complete loss of external differentiation into testis and ovary. The macroscopic and histological details of the various stages of gonad development are given in *Table 2.*

The juvenile scallop The breeding cycle of the juvenile scallop begins with those scallops which have just spawned for the first time in August or September. The gonads began to show signs of recovery externally in October or November (*Table 4*) although production of gonia had occurred before this. Since most gonads had spawned completely and lost their differentiation into testis and ovary, it was possible to distinguish stages I (spent-recovering) and II (differentiated), which cannot be distinguished in adult gonads owing to their retention of more genital products and, consequently, of their differentiation. Recovery continued throughout the winter, stages III and IV predominating from January to March. The third growth ring was laid down on the shell in the spring. Stage V gonads were commonest in April and May, and by June or July most juvenile scallops had full (stage VI) gonads. The main spawning occurred in the second half of August in 1951 and between 5 and 11 September in 1952 (*Table 4B*) and most scallops of this age took part. Spawning was almost complete in most individuals, but more sexual products were retained than in the first (virgin) spawning, and the gonads usually retained their differentiation. In addition, some juvenile scallops released a very small proportion of their gametes between 9 and 17 July 1952 (*Table 4B*).

The adult scallop The principal spawning of adult scallops, like that of juveniles, was found to occur in August or September. At the beginning of each cycle were found mostly spent and a few

Table 4 Percentage of juvenile scallops at each stage of maturity (Mason, 1958a)

(A) Monthly percentage
(All samples in each month combined)

October 1950–September 1951
Gonad stages

Month	*I*	*II*	*III*	*IV*	*V*	*VI*	*VII*	*No. of gonads*
Oct	–	–	–	–	–	–	100.0	6
Nov	25.0	8.3	16.7	8.3	–	–	41.7	12
Dec	–	25.0	–	50.0	–	–	25.0	4
Jan	17.6	19.6	5.9	33.3	7.8	–	15.7	51
Feb	17.0	5.4	19.7	39.5	4.8	–	13.6	147
Mar	11.9	6.8	13.7	36.3	15.9	7.9	7.9	278
Apr	2.3	4.7	7.0	30.2	14.0	41.9	–	43
May	–	0.8	2.5	27.5	34.2	27.9	7.1	120
June	–	–	–	0.8	27.7	68.1	3.4	177
July	–	–	–	–	10.3	88.5	1.3	39
Aug	–	–	–	–	21.2	51.5	27.3	33
Sep	–	–	12.3	36.3	0.9	12.3	38.2	106

September 1951–October 19 2
Gonad stages

Month	*I*	*II*	*III*	*IV*	*V*	*VI*	*VII*	*No. of gonads*
Sep	–	–	16.7	22.2	–	–	61.1	18
Oct	–	14.3	21.4	14.3	–	–	50.0	14
Nov	2.3	4.5	25.0	13.6	2.3	–	52.3	44
Dec	–	–	–	37.5	–	–	62.5	8
Jan	2.4	11.9	19.0	33.3	13.1	1.2	19.0	42
Feb	6.0	1.2	20.2	46.4	16.7	7.1	2.4	84
Mar	–	–	11.1	50.0	22.2	16.7	–	18
Apr	1.1	1.4	2.8	37.9	43.3	13.2	0.3	178
May	–	–	1.3	10.2	51.6	36.8	–	152
June	–	–	–	1.4	56.4	40.0	2.1	70
July	–	–	–	6.7	15.6	62.2	15.6	90
Aug	–	–	–	6.8	12.8	65.5	14.9	74
Sep	–	–	–	4.1	8.2	58.2	29.6	49
Oct	–	–	34.4	37.5	–	9.4	18.8	32

(B) Spawning months
(Samples arranged to show dates between which spawning occurred)

Gonad stages

1951	*I*	*II*	*III*	*IV*	*V*	*VI*	*VII*	*No. of gonads*
15, 18 Aug	–	–	–	–	25.0	75.0	–	4
31 Aug	–	–	–	–	20.7	48.3	31.0	29
5, 12 Sep	–	–	–	12.0	–	38.0	50.0	25
21, 28 Sep	–	–	16.0	43.8	1.2	4.3	34.6	81

Gonad stages

1952	*I*	*II*	*III*	*IV*	*V*	*VI*	*VII*	*No. of gonads*
9 July	–	–	–	–	26.3	73.7	–	38
17 July	–	–	–	9.1	–	45.5	45.5	11
21, 29 July	–	–	–	12.2	9.8	56.1	22.0	41
5 Sep	–	–	–	5.7	11.4	72.9	10.0	35
11 Sep	–	–	–	–	–	21.4	78.6	14

partially spent gonads resulting from the spawning of juvenile and adult scallops (*Table 5*). Recovery soon began, and, owing to the larger numbers of reproductive cells retained by juveniles and adults after spawning, the first recognizable stage in recovery was stage III (recovering). Recovery continued throughout the winter months, stages III and IV predominating in October and November, and stages IV and V in December. Half-full (stage V) gonads were abundant in the late winter, and by March or April most adult scallops had full (stage VI) gonads. The full gonads, however, still had considerable numbers of spermatocytes and growing oocytes in the follicles, so that, although a mass-spawning occurred in the spring, it was only partial. In 1951 this spawning occurred between 4 and 11 May, and in 1952 between 12 and 15 April (*Table 5B*). The spring spawning was followed at once by recovery, and, since only a proportion of the gonad contents had been shed, the resulting partially spent gonad resembled somewhat the stage III gonad, so that the first recognizable stage of recovery was stage IV. Recovery continued throughout the summer, through stages IV and V, so that by July most adult scallops again had full (stage VI) gonads, this time with few spermatocytes and growing oocytes. These then took part in another mass spawning, which resulted, in most individuals, in the production of almost completely spent gonads. This spawning occurred at approximately the same time as that of the juveniles, *ie* late in August (actually between the 18th and 23rd) in 1951 and between 5 and 11 September in 1952 (*Table 5*). In addition, between 9 and 17 July 1952, again at the same time as the juveniles, a few adult scallops spawned slightly, and recovered quickly to take part in the main September spawning. In 1951, a sample taken on 15 August showed that a similar slight spawning had occurred in a few scallops between 24 July and 15 August, again before the main spawning.

The majority of adult scallops, then, spawn together twice during each annual breeding cycle, partially in April or May ('spring' spawning), and more completely in late August or September ('autumn' spawning). There is also a minor 'summer' spawning in July or early August. Virgin and juvenile scallops differ from adults in having only one major spawning, in autumn, though juveniles show some evidence of a minor summer spawning. The autumn spawning is therefore by far the most important in terms

Table 5 Percentage of adult scallops at each stage of maturity (Mason, 1958a)

(A) Monthly percentage
(All samples in each month combined)

October 1950–September 1951

Month	*III*	*IV*	*V*	*VI*	*VII*	*No. of gonads*
Oct	31.3	34.3	–	–	34.3	67
Nov	26.3	62.5	6.9	–	4.4	160
Dec	8.5	74.5	17.0	–	–	47
Jan	1.2	36.8	54.0	8.0	–	201
Feb	2.0	14.3	59.2	24.5	–	98
Mar	0.3	4.4	32.9	62.2	0.1	343
Apr	0.4	2.0	6.5	90.4	0.6	245
May	–	10.1	4.1	39.0	46.9	318
June	–	2.9	76.9	10.8	9.4	346
July	–	–	10.0	85.7	4.3	70
Aug	–	0.7	23.4	25.2	50.7	141
Sep	5.9	25.8	5.9	14.1	48.3	145

August 1951–October 1952

Month	*III*	*IV*	*V*	*VI*	*VII*	*No. of gonads*
Aug	–	–	12.4	27.0	60.6	113
Sep	9.4	30.2	4.0	13.9	42.5	260
Oct	23.0	63.1	3.4	4.9	.5.5	293
Nov	9.2	53.9	31.4	4.5	1.0	191
Dec	1.8	33.5	40.4	22.0	2.3	109
Jan	1.0	15.1	55.4	28.3	0.2	288
Feb	–	3.7	49.8	45.7	0.8	246
Mar	–	1.5	17.6	79.9	1.0	259
Apr	–	21.0	7.1	59.3	12.5	554
May	–	5.1	79.6	15.2	0.2	445
June	–	0.2	52.6	46.5	0.7	227
July	–	5.2	10.0	76.2	8.6	296
Aug	–	7.1	10.7	71.1	11.1	140
Sep	–	–	6.8	52.1	41.1	96
Oct	43.5	42.7	–	2.4	11.3	62

(B) Spawning months
(Samples arranged to show dates between which spawning occurred)

Gonad stages

1951	*III*	*IV*	*V*	*VI*	*VII*	*No. of gonads*
4 May	–	–	3.3	93.3	3.3	60
11 May	–	5.2	3.4	29.3	62.1	116
18 May	–	18.3	4.9	23.9	52.8	142
17, 24 July	–	–	10.0	85.7	4.3	70
15, 18 Aug	–	1.8	45.5	34.5	18.2	55
23, 31 Aug	–	–	9.3	19.2	71.5	86

Gonad stages

1952	*III*	*IV*	*V*	*VI*	*VII*	*No. of gonads*
1-12 Apr	–	0.3	6.6	92.5	0.5	286
15, 17 Apr	–	–	3.6	41.6	54.8	83
25 Apr	–	62.4	9.4	15.9	12.2	185
9 July	–	–	13.2	86.8	–	148
17-29 July	–	10.5	6.8	65.5	17.2	148
5 Sep	–	–	11.0	83.1	5.9	59
11 Sep	–	–	–	2.7	97.3	37

of numbers of gametes released.

A similar pattern of spawning with two mass spawnings or peaks of spawning activity each year, a partial or minor one in the spring and a complete or major one in the autumn, has been reported in other areas. These include southwest Ireland (Gibson, 1956), Northern Ireland (Strangford Lough – Stanley, 1967), northwest Scotland (Loch Torridon – Mason, 1969) and the Clyde Sea Area (Cumbrae – Comely, 1974). My own unpublished observations suggest a similar situation in the Clyde and off the southwest of Scotland outside the Clyde, though the indications are that spawning is somewhat later outside than inside the Clyde. At Holyhead, North Wales, Baird (1966) found two spawnings each year but in four years out of five the spring was the more productive spawning. Buestel, Dao and Lemarié (1979) found that in the English Channel spat (see later) settled in a number of waves in July to September and inferred that spawning did not show a few peaks but was more continuous (see also Franklin, *et al,* 1980).

Spawning

The spawning of *Pecten maximus* was observed on several occasions in the laboratory. The genital products were passed out through the two main ducts, through the kidneys, and into the mantle cavity, whence they were emitted in a cloud through the exhalant opening of the shell. No violent flapping of the shell valves occurred, but they opened and closed gently at intervals of 1 to 2 minutes. The eggs settled and formed an orange layer on the bottom of the container, while the sperm became dispersed in the water and made it cloudy. Both kidneys became filled with either eggs or sperm. Eggs and sperm were not shed at the same time, but usually within a few hours of one another, sometimes the eggs and sometimes the sperm being shed first. In about 4% of cases one or other of the two parts of the gonad remained unspawned.

On the only occasion on which the act of spawning was observed from the beginning, the ovary spawned in 45 minutes and was followed at once by the testis, which became spent in two hours. On several occasions both parts of the gonad spawned overnight.

Though some other writers (Franklin, *et al* 1980; Pickett, 1978) have stated that the sperm are normally shed first, Comely

(1972) confirmed my finding that the products of either sex can be extruded first. Comely also found that spawning in aquarium tanks mostly occurred either late in the evening or at night.

The running gonad has a patchy appearance, having dull areas which have shed all their ripe products and bright areas which still contain them.

Sections of a running testis (*Plate 19*) show follicles in varying stages of spending; most follicles are partly empty, having spaces in their centres, but, while these spaces contain many free spermatozoa, some spermatozoa are still arranged radially, as in the full gonad, with their tails pointing inwards. The ducts are wide, and their lumina contain spermatozoa. The duct walls contain a secretion which stains purple in haematoxylin, which is thought to facilitate passage of the genital products.

Sections of a running ovary (*Plate 19*) show that some follicles are in a spent condition, and contain the remains of oocytes, while others are still full, and in some of these are seen ova which have undergone or are undergoing maturation. In these ova, the germinal vesicle has broken down, the cytoplasm is evenly distributed,

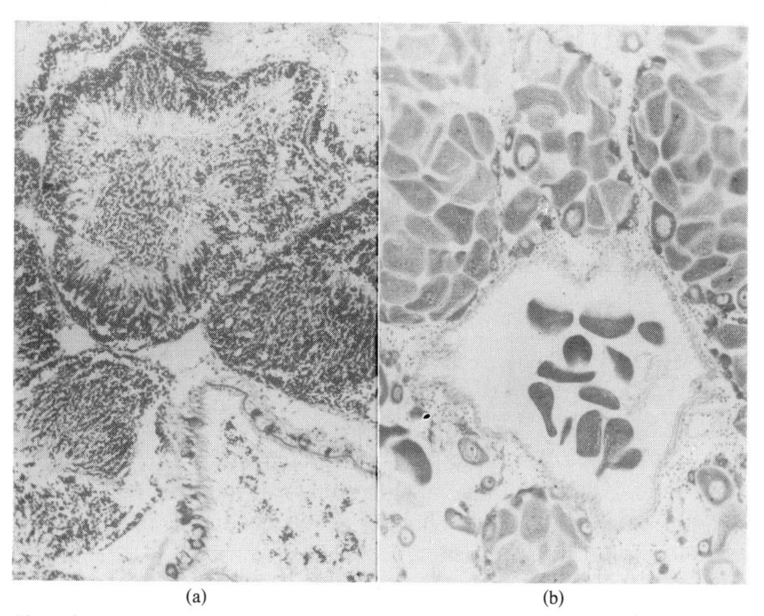

(a) (b)

Plate 19 Photomicrographs of transverse sections through running gonad, (a) testis and (b) ovary (both magnified x 94) (*Mason, 1958a*)

and a spindle with chromosomes can sometimes be seen. The ovum is still polygonal, and possesses a conspicuous membrane. Other follicles have shed a few ova, and have a few loose in their lumina. The ducts are wider and contain mature ova, and a purple-staining secretion is sometimes present in their walls.

It was found that ripe oocytes, cut from the full gonad and placed in sea water, quickly became spherical, lost the germinal vesicle, and could then be readily fertilized artificially. Eggs passed naturally from the gonad into the surrounding water behave similarly. These facts suggest that contact with sea water might initiate the onset of maturation. It is possible that the eggs are often passed out of the ovary before maturation, since, in an ovary which has commenced to spawn, many fully-grown oocytes still retain their germinal vesicles, and also that contact with sea water in the mantle cavity results in the dissolution of the nuclear membrane. The genital ducts in the running gonad are so wide that water from the mantle cavity and kidneys might conceivably pass along them to the follicles and cause the onset of maturation in some oocytes before they leave the follicle. Coe (1933) found such a state of affairs in *Teredo*, with sea water causing the initiation of maturation.

Tang (1941) stated that *Pecten* off Port Erin began to spawn in numbers when the water temperature reached 10°C, but this is contradicted by my study. Of the six spawnings which occurred in 1951 and 1952 it was possible in five cases to state, to within a few days, when spawning had occurred (the exception was the small summer spawning of 1951, when insufficient samples could be taken). Spawning occurred at several temperatures between 7.2 and 13.7°C, viz spring 1951 – 7.2°C, autumn 1951 – 13.7°C, spring 1952 – 8.1°C, summer 1952 – 12.7°C, autumn 1952 – 13.5°C.

Fertilization

Fertilization is external, and, at the times of the mass spawning, there are relatively high concentrations of genital products in the sea, greatly increasing the chances of successful fertilization.

In the laboratory, artificial fertilization of eggs cut from the gonad was readily carried out. In nature, where each individual sheds its two types of genital products separately into the sea, cross-fertilization must be the general rule. Experimentally, however, cross- and self-fertilization were obtained with equal facility. On one occasion a rough count showed that about 80% fertilization was obtained, though usually the percentage was much lower than this. Better results were obtained using ripe gametes which had been shed normally by the scallop.

Normally only mature eggs from a full gonad are capable of being fertilized. However, sperm from stage IV, V and VI gonads, as well as residual sperm from partially spent gonads, are capable of fertilizing them.

Larvae and spat

Our knowledge of the larval stage of the scallop is rather sketchy and is based on a combination of direct observation in the sea and the behaviour of larvae reared under hatchery conditions. Comely (1972), Gruffydd and Beaumont (1972), Buestel, Arzel, Cornillet and Dao (1978) and Sasaki (1979) succeeded in fertilizing eggs and rearing the resulting larvae. The shelled veliger appeared two days after fertilization and the crawling pediveliger stage was reached two days before metamorphosis. At 16°C metamorphosis at 250μm occurred 33–40 days and at 18°C three weeks after fertilization. In nature the time may be 3–4 weeks (Franklin *et al*, 1980). After a short time near the sea bed, the veliger larvae rise to the surface layers. As the time for metamorphosis approaches the larvae descend to the sea bed.

Plankton samples taken off Bay Fine, Isle of Man (*Fig 18*), just below the surface, using a fine townet, 129 meshes to the inch, at intervals of one to two weeks, during the period 28 April to 24 December 1952 showed three main peaks of abundance, two large ones on 28 April and 5 September, and a smaller one on 29 July (*Fig 19*). The first large peak would probably be composed of larvae arising from the spring spawning, the small peak of larvae from the small summer partial spawning and the second large peak of larvae from the autumn spawning. Since the mass spawning of scallops in the population studied did not occur until between 5 and 11 September, the larvae found on 5 September probably originated from the spawn of a different group of scallops in the vicinity, which had spawned somewhat earlier.

The large numbers of larvae in the plankton haul of 28 April show that the spring spawning resulted in successful production of larvae. When spawning was recorded, between 12 and 15 April,

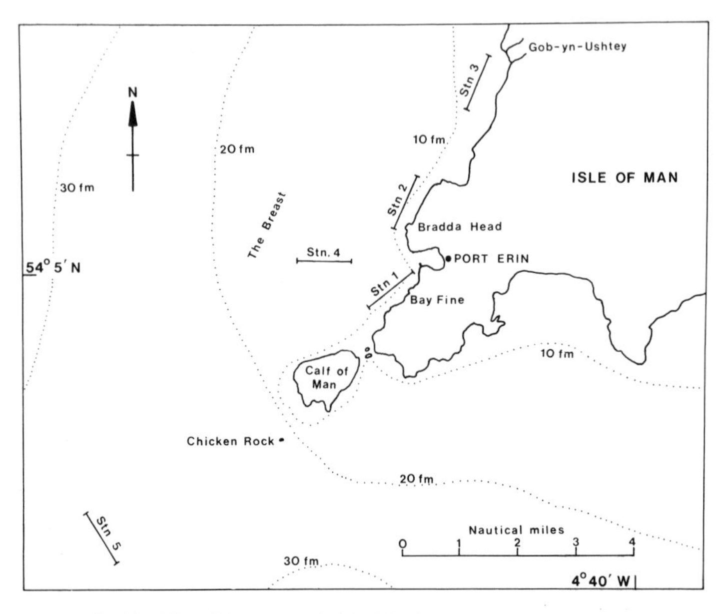

Fig 18 Map of the south end of the Isle of Man showing dredge stations

the temperature was 8.2°C, indicating that a temperature of 10°C is not minimal for the production of larvae, as was thought by Tang (1941).

The sampling was not adequate to show the exact times and composition of the peaks in *Fig 19*, this being especially so in the September peak, when there was a gap of 25 days after the sample of 5 September. The September peak would be expected to be much higher than the April peak, owing to the much greater amount of spawn shed in the autumn than in the spring spawning.

Various workers have commented on the scarcity of recently settled spat as well as young (0+ to 2+) scallops on commercial beds and it has been postulated that spat settle on grounds separate from the adults, possibly inshore. However, some of the commercial beds, especially in the English Channel (Baird, 1952; Franklin *et al,* 1980), are several miles offshore, and it is unlikely that the scallop with its limited powers of swimming (Soemodihardjo, 1974; Minchin, 1978a) could undertake such long migrations. It seems much more likely that the spat and young would occur on the same grounds as the adults. However, in two years of intensive dredging off the Isle of Man, I (Mason, 1958a) found

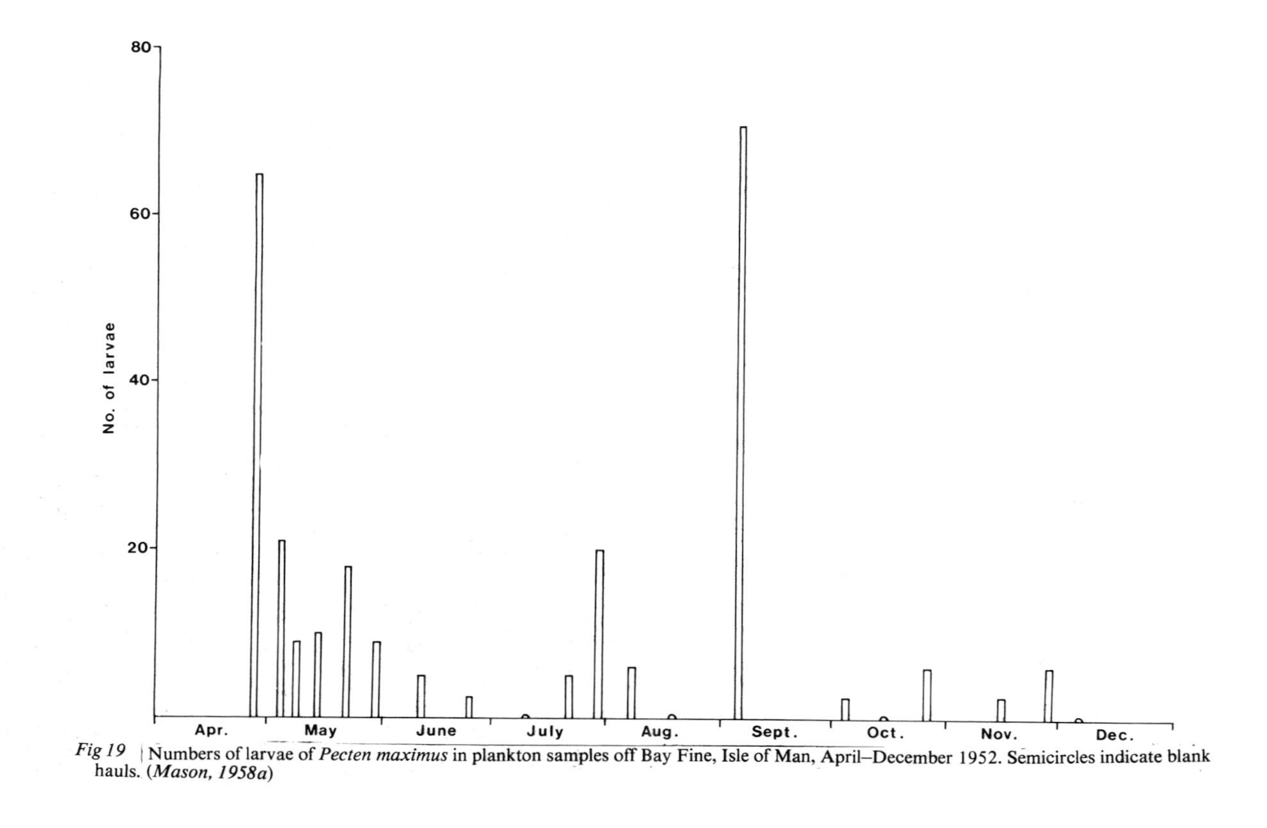

Fig 19 Numbers of larvae of *Pecten maximus* in plankton samples off Bay Fine, Isle of Man, April–December 1952. Semicircles indicate blank hauls. (*Mason, 1958a*)

only two, 3.0 and 3.3mm long, attached to the alga *Desmarestia*. Eggleston (1962), by carefully sifting through bottom samples taken off the Isle of Man, found 48 specimens ranging from 0.8 to 6.6mm in height, some attached to hydroids and polyzoans and some unattached. Elmhirst (1945) and subsequently Mason and Drinkwater (1978) have found fair numbers 2–11mm long attached to *Laminaria saccharina* off the west coast of Scotland in depths of 6–25m. It appears that they can settle on a variety of substrata above the sea bed, including man-made fibres, a property which is providing the basis of attempts to cultivate scallops (see *Chapter 11*), though spat of *Pecten maximus* are usually far outnumbered by those of *Chlamys opercularis*. Paul (1978) and Brand, Paul and Hoogesteger (1980) suggested that both *P maximus* and *C opercularis* first settle on a variety of substrata, mostly bryozoan and hydroids, which are erect, raised above the sea bed and silt-free. Indeed, settlement experiments by Comely (1972) revealed no substratum preference, the larvae attaching themselves to the lowest possible part of the containing vessel, whatever its nature.

At this stage a byssus can be secreted by the byssal gland in the foot through the byssal notch, but this can be broken and reformed at will, and the animal can move, either crawling by means of its foot or flapping its shell valves and swimming (Minchin, 1978a).

Recent work by Minchin (1978a) in the Republic of Ireland has suggested a reason why so few scallop spat have been found. In the course of diving observations he found spat from 2–15mm attached by the byssus to various algae and hydroids. He also, however, found spat 4–7mm unrecessed on the sea bed and others 7–25mm attached to *Lithothamnion*. He also found many spat as small as 6–10mm recessed in the substratum in a depth of 13m, and the method of recessing was as described in adults by Baird (1958). It appears likely, therefore, that spat are present on the same grounds as adults, but have hitherto remained undetected because of the small size at which they became recessed. Some of Minchin's recessed spat also had a byssus, by which they were attached to shell fragments. *P maximus*, however, generally loses its byssus much earlier than does *C opercularis*, and few scallops larger than 14–15mm have been seen attached (Franklin *et al*, 1980; Mason, 1969). However, on one occasion my attention was drawn by a Port Erin lobster fisherman to two young scallops approximately 30mm long which were attached to the twine cover of one of his lobster creels. *Chlamys opercularis*, whilst it is usually free-living after attaining some 15–20mm, retains the ability to secrete a byssus until it is at least 65mm long.

Even the two recently settled individuals as small as 3.0 and 3.5mm found off Port Erin were easily recognized, since the shape of the young shell can be deduced from the concentric striae on the adult shell (Mason, 1957). The shell was transparent and almost colourless, with a conspicuous byssal notch, and both shell valves were convex. Larger specimens, some 5–25mm long, possess a left valve which is concave except for a small convex area near the umbo, while the right valve is convex. Ribs begin to appear 8–10mm from the umbo, and the shell then begins to take on the adult form, with a flat left valve (*Plate 20*). The byssal notch has almost disappeared after about two years, when the scallop is about 50–55mm long.

The breeding of *Chlamys opercularis*

The gonad of *Chlamys opercularis* is very similar in form to that of *Pecten maximus* though in the queen the loop of the alimentary canal does not penetrate far, if at all, into the female part (Fullarton, 1890; Dakin, 1909). The gonad of *C opercularis* is plumper and when full is almost circular in section. The female part is usually a deeper red than that of *P maximus*.

A number of workers have studied the breeding cycle of queens in Manx waters. Gametogenesis begins early and has been observed in animals as small as 9mm (Aravindakshan, 1955), and the smallest with a pinky tint was 21mm high (Soemodihardjo, 1974). The development was studied by dividing it into five arbitrary stages based on those of Mason (1958a). Queens attain sexual maturity and spawn for the first time when barely a year old, in the autumn, though the gonads are so small that this spawning probably contributes little. The second spawning occurs a year later, again in the autumn (Aravindakshan, 1955). Subsequently, there are three mass spawnings per year, an intense autumn spawning (August–October) and two minor partial ones, one in winter (January–February) and one in spring or early summer (May–July), the spring/summer spawning being followed by rapid recovery (Soemodihardjo, 1974; Paul, 1978).

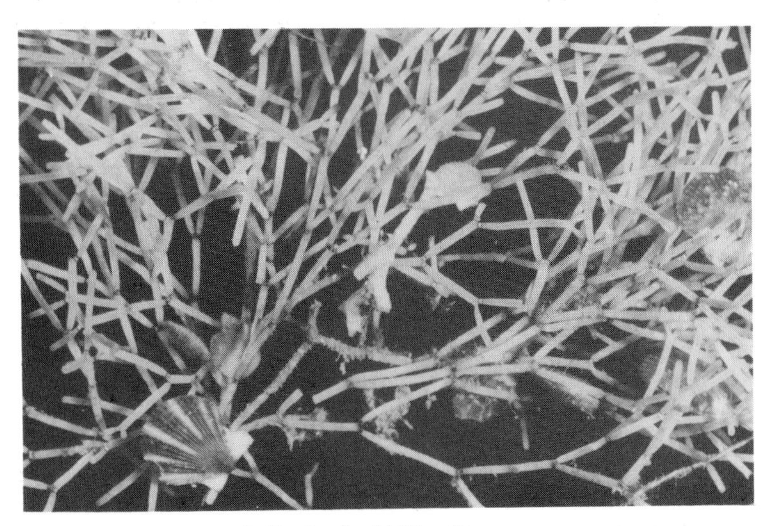

Plate 20 (a) Spat of queen (left) and scallop (right), *ca* 3mm
(b) Queen spat, up to *ca* 7.5mm, attached to *Cellaria* (Bryozoa) (*By courtesy of A R Brand*)

The pattern of spawning, like that of *Pecten maximus*, appears to vary from place to place (Broom, 1976). Thus Amirthalingam (1928) found that at Plymouth (SW England) the main spawning season extended from January to June, with a number of spawnings coinciding with full moons though smaller spawnings occurred at new moon in July and October. Broom and Mason (1978) found three major spawnings, spring, autumn and winter, in queens kept in cages at Portsmouth on the English Channel. Taylor and Venn (1979) found a main spawning in July – August in scallops in the Clyde sea area, and a small spawning (possibly a partial spawning by only a proportion of the population) in the winter. Fullarton (1890) found that peak spawning in the Firth of Forth occurred in July and August.

Gametogenesis occurs as in *Pecten maximus* and the appearance of the follicles is similar in sections to those of *P maximus*. Again the polyhedral oocytes soon become spherical when released into the sea, and have a diameter of 70μm (Fullarton, 1890). The sperm has an ovate head drawn out at the apex, 1.5μm long, and a tail 50μm long (Aravindakshan, 1955). According to Aravindakshan (1955) the sperm are always shed before the eggs, though Fullarton (1890) said either could be shed first. Fertilization is external, and maturation is completed after entry of the sperm (Broom, 1976). The planktonic larvae develop to a size of 250–280μm (Jørgensen, 1946; Rees, 1950) probably in 20–30 days (Paul, 1978). Again the larvae settle on a variety of types of substratum, including hydroids, bryozoans (especially *Cellaria, Plate 20*) and algae, and possess the ability to delay metamorphosis until a suitable one is found. Generally recently-settled spat of *C opercularis* are many times more abundant than those of *P maximus* (Mason, 1969; Ventilla, 1977). This is almost certainly due to the much greater abundance of adults of *C opercularis* than of *P maximus*. While a density of one per 10m^2 to one per 5m^2 is considered a good commercial density in *Pecten maximus, Chlamys* is found in much greater densities. Divers have found queens three or four deep (500–1,000 per m^2) on the sea bed over a considerable area in the Firth of Clyde.

The queen spat have generally left the original substratum and settled on the sea bed by the time they have reached a size of 15–20mm, though they retain the ability to secrete a byssus and attach themselves to shells and other objects much longer than the scallop. Animals as large as 40mm have been seen crawling by protruding the foot, secreting a byssus and drawing up the body to the foot (Paul, 1978); the byssus is then broken and the process repeated. Animals as large as 65–70mm have been seen to secrete a byssus in laboratory conditions (Soemodihardjo, 1974). The byssal notch is retained throughout life.

Discussion of breeding

In the system of classification used by Coe (1945), both *Pecten maximus* and *Chlamys opercularis* are 'functional hermaphrodites', in which reproductive cells of both sexes are produced at the same time but are not usually discharged together. In *P maximus*, the products of either sex may be discharged first, but they are discharged within a short time of each other. The same may be true of *C opercularis*, for although Aravindakshan (1955) said that the sperm are always discharged first, Fullarton (1890) found that the products of either sex could be spawned first.

Mason (1958a) showed that in the virgin gonad of *P maximus* at least there is an early tendency, not towards protandry, but towards protogyny, in that oocytes are produced before spermatocytes (*Table 2*). Later, however, this tendency is reversed, and spermatogenesis so far outstrips oogenesis that, as in *C opercularis*, the male follicle contains some spermatozoa while the female follicle still contains nothing bigger than partially-grown oocytes. Similarly, after spawning, spermatogenesis occurs at first more rapidly than oogenesis and again many sperm are present long before the oocytes are fully grown. These sperm from the earlier gonad stages are physiologically ripe and active and capable of fertilizing ripe eggs which, however, occur only in full gonads. The sperm are stored in the testis until the gonad is full and discharged separately from the eggs but within a short time of them.

The spawning cycles of both *Pecten maximus* and *Chlamys opercularis* are closely tied to the seasonal changes in biochemical composition and weight of the adductor muscle and other tissues. Studies of biochemical composition in *P maximus* include those of Mason (1959b), Stanley (1967) and Comely (1974) and in *C opercularis* those of Soemodihardjo (1974) and Taylor and Venn (1979). The cycles are similar in the two species, and the following represents a consensus of the views of various authors.

The dry weight of the adductor muscle is lowest in spring but increases throughout the spring and summer, when food is abundant and temperature is increasing, and reaches a peak in the autumn. The increase is attributable to an increase in carbohydrate and protein content. The adductor muscle then becomes increasingly watery and its dry weight decreases throughout the autumn and winter, when food is scarce and the temperature low, to the spring minimum. This is due to chemical breakdown of cellular material for gametogenesis and gonad development, which occur steadily throughout the autumn and winter after the summer–autumn spawning, and to metabolic maintenance. The quick muscle is believed to be the chief nutrient store. During the adverse conditions in winter these stores are sufficient to maintain metabolism and gonad development but are unable to maintain growth, which therefore ceases until the more favourable conditions return in the spring (see *Chapter 8* and Orton, 1928; Mason, 1957; Broom and Mason, 1978).

For spawning to occur, a proportion of the gametes must be ripe, probably a much higher proportion before a complete than before a partial spawning. However, the presence of ripe products does not necessarily result in immediate spawning – ripe products can be stored for a considerable time, sperm probably several months. The actual factor which triggers off spawning has been the subject of conjecture. The attainment of a particular temperature can be ruled out since spawning in both species has been observed over a wide range of temperature, and both when the temperature is rising and when it is falling. Similarly the abundance of food is excluded. Experimentally spawning has been initiated by shock, *eg* by exposure to air followed by re-immersion, and by rapid variation in temperature, though such factors can scarcely be responsible in nature. Spawning of both species has been associated with full and new moon, or spring tides (Amirthalingam, 1928; Mason, 1958b). Comely (1972) has speculated that the act of dredging could cause a shock sufficient to initiate spawning. Whatever the initial trigger, it appears likely that the spawning of one individual can trigger it off in others.

The success of a spawning can thus be affected by environmental conditions during gametogenesis, which influence the numbers of ripe gametes. Conditions after fertilization can affect the survival of larvae and spat and so influence the success of a brood.

8
Age and growth

Age and growth of the two species *Pecten maximus* and *Chlamys opercularis* will be treated separately owing to the differences between the two species.

Age and growth in *Pecten maximus*

Age analysis

This description is based largely on my own study of growth of scallops at Port Erin (Mason, 1957). The material was obtained from commercial and small mesh dredge hauls made at five places off the southwest of the Isle of Man (*Fig 18*).

Various measurements were made on the scallops caught. The overall length (anterior-posterior axis), breadth (dorsoventral axis), and thickness (lateral axis) were measured on a specially designed measuring board. The overall length and breadth are in reality the length and breadth of the rounded, or right, valve of the shell which overlaps the flat, or left, valve. Dividers were used to measure the distance from the umbo to the edge of the flat valve, and to each of the annual growth rings (see below) on the flat valve along the dorsoventral axis (*Fig 1*). All measurements were to the nearest millimetre below the value shown on the scale.

The most convenient dimension is the overall length, which is easiest to measure and the least ambiguous to use in legislation since it is the maximum dimension in *Pecten maximus* at all sizes and in *Chlamys opercularis* of commercial size. Some authors, (*eg* Soemodihardjo, 1974; Taylor and Venn, 1978) have preferred the term 'height' to 'breadth', and while this is anomatically correct, I feel that 'breadth' is likely to be less confusing to fishermen and legislators.

The growth rings The shell of *Pecten maximus* grows as the mantle secretes layers, or lamellae, on its inner surface, each of which emerges at the edge of the shell from under the previous one. This results in the formation on the outer surface of concentric striae or ridges, mostly 0.1–0.3mm apart, which are prominent and raised over most of the shell but less so in a slightly concave area within 20–25mm of the umbo (*Plate 21*). Each stria therefore represents the edge of the shell at the time of its deposition. The striae tend to become worn on the flat valve, owing to abrasion by the substratum, and because of this they are best studied on the flat upper valve.

The shell also bears distinct concentric growth rings, which are white in colour and translucent (*Plate 1*). They show more clearly on the reddish-brown flat valve than on the white round valve. They occur regularly and in approximately the same position on most shells.

In scallops taken in 13–16fm (23.8–29.3m) off Port Erin, Isle of Man (Mason, 1957), the rings were found to be 0.5mm or less wide, and made up of striae which are crowded together about 0.05mm apart and less prominent than elsewhere (*Plate 21*).

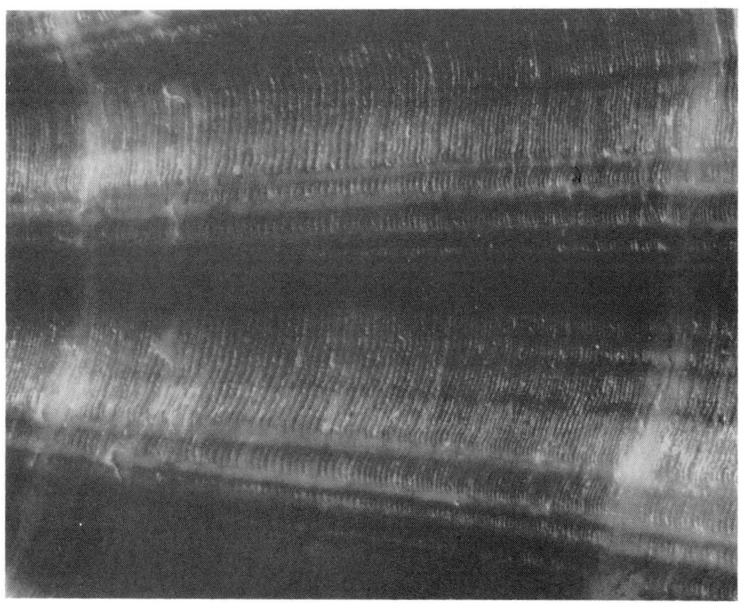

Plate 21 Flat valve of the scallop showing the arrangement of striae within and between successive annual rings

Immediately outside a ring, on the side away from the umbo, the striae become more raised, assume a pale brown colour, and become farther (about 0.3mm) apart. After a few millimetres the striae take on a darker hue, and become gradually closer together until, just before the next ring, they are about 0.1mm apart. This description applies to the shell between the second and third rings from the umbo. Farther from the umbo, both the growth rings and the striae are somewhat closer together, but a similar series of changes is seen. The distance apart of the striae reflects seasonal differences in growth rate.

An examination of the edge of the flat valve of the shell throughout the year showed that only one ring is laid down each year, in the spring. The growth ring is narrow and is laid down slowly but its deposition is followed by increasingly rapid growth, so that any growth ring near the edge of the shell has very recently been laid down. In each month the number of scallops with fewer than five growth rings which had a ring within 3mm of the edge of the flat valve was noted (*Fig 20*). The somewhat low percentages are due to the long period of time (March–May) during which rings are laid down, so that some scallops have acquired more than 3mm of new growth before others have commenced to lay down their ring.

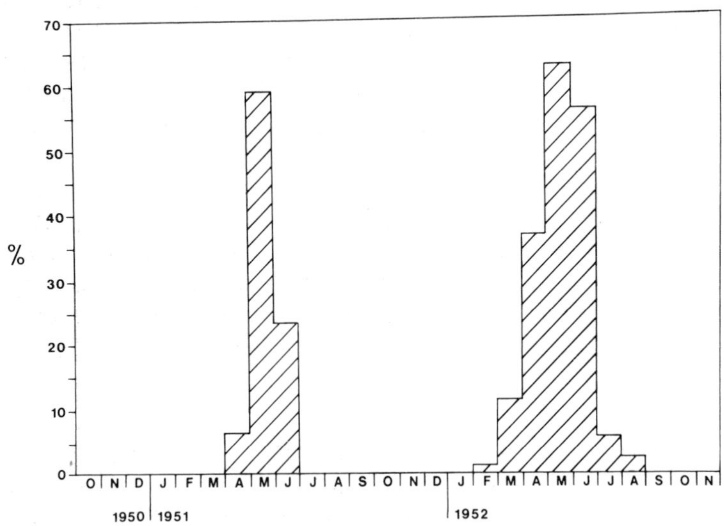

Fig 20 Percentages of *Pecten maximus* with fewer than five growth rings which had a growth ring within 3mm of the edge of the shell, Port Erin, Isle of Man. (*Mason, 1957*)

Evidence that only one ring was laid down each year was afforded also by tagging experiment. Manx scallops were tagged by a small plastic disc wired to an auricle. Six living tagged scallops were recaptured which had spent one spring in the sea, and each of these had one more ring than when it was released. Of 37 scallops recaptured before they had spent a spring in the sea, none had acquired a further ring. Occasionally a disturbance ring was found on these scallops where the edge of the shell was at the time of tagging, being probably caused by a retraction of the mantle edge away from the edge of the shell during tagging. Such a ring differs from an annual ring in that it has no small, crowded striae, and that the striae on either side of it are equally spaced. Such false rings also occur occasionally in nature, and indeed virtually every scallop observed in the Clyde Sea Area formed a disturbance ring in the summer of 1960 – the cause was not known, but must have been a widespread environmental factor. Such rings on individual shells might be caused by capture and return to the sea or even contact with fishing gear, both of which cause closure of the shell and probably retraction of the mantle from the edge of the shell.

Gibson (1956), by means of a similar tagging experiment, obtained evidence of the annual nature of the growth rings on the shell of *P maximus* in southern Irish waters. The annual nature of the rings has in fact now been demonstrated or assumed in scallops in all parts of the British Isles (*eg* Scotland – Mason and Drinkwater, 1969, 1973, 1974, 1975, 1976; Northern Ireland – Stanley, 1967; North Wales – Baird, 1966; English Channel – Pickett, 1978; Franklin *et al,* 1980).

The area of shell between two successive annual growth rings, or between the umbo and the first ring, is known as a growth band. A scallop is aged by the number of growth rings and the presence or absence of new growth outside the outermost ring. Thus 5+ indicates that the shell has five rings, with new growth outside the fifth, while 4 indicates that the fourth ring has just been laid down at the edge of the shell.

Length of life Tang (1941) recorded a scallop, captured off Port Erin, which had 22 growth rings; the oldest one I caught had 18 rings. Rings beyond the ninth or tenth are, however, difficult to distinguish, and some uncertainty is attached to these ages. The biggest I have seen recorded in the literature was from Cork (Southern Ireland); it was 214mm long and had nine growth rings (Minchin, 1978b).

Growth

The first growth band When the size frequency of the first growth band of Port Erin scallops was plotted, a bimodal distribution was obtained (*Fig 21*). The modal values were 19 and 39mm respectively, and the curve was divided arbitrarily at 28mm, giving two groups of scallops, a majority with small, and a minority with large, first growth bands. Of 4,379 scallops examined, 4,049 or 92.5% had small, and 330 or 7.5% had large, first growth bands.

The two types of first growth band can be correlated with the breeding cycle of the scallop. As previously indicated *P maximus* in Manx waters has two main spawnings each year, a spring spawning in April or May and an autumn spawning in late August or September, while there is a small summer spawning in July or early August. Growth of the juvenile and adult scallop ceases in December, and the resumption of growth in spring results in the appearance of the growth ring (see later). If this is true of the O-group also, spring spawned scallops would thus grow for a greater length of time before the first cessation of growth than would

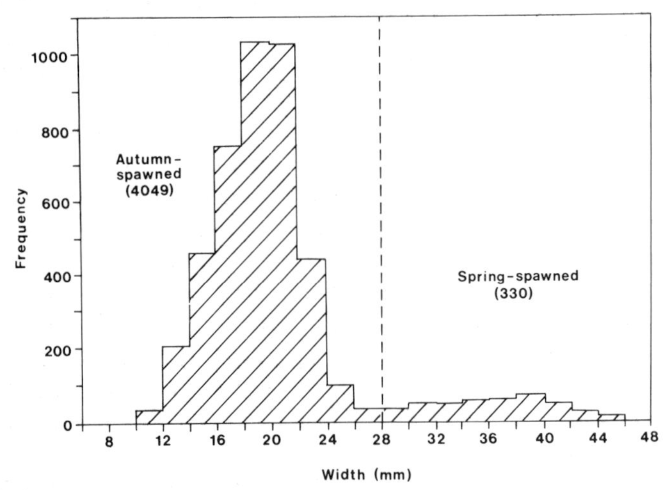

Fig 21 Width frequency of the first growth band of *Pecten maximus*, Port Erin, Isle of Man. (*Mason, 1957*)

autumn spawned individuals. It is suggested that most of the scallops forming the minor group, those with large first growth bands, arise from the spring spawning, and that more of those forming the major group, those with small first growth bands, arise from the autumn spawning, while a few of the latter probably arise fom the small summer spawning.

Although there are many possible factors which can influence the success of a brood, the difference in the number of scallops constituting the two groups can, to some extent, be accounted for by the amount of spawn released in each spawning. Only those which have just deposited, or are about to deposit, their fourth or any subsequent growth ring, take part in the spring spawning, and their gonads become only partially spent. On the other hand, mature scallops of all ages (those with two or more growth rings) take part in the autumn spawning, and the gonads most often become completely spent. Thus more gametes are shed in the autumn than in the spring spawning. Scallops with large first growth bands are therefore called 'spring spawned' and those with small first growth bands 'autumn spawned'. Gibson (1956) and Stanley (1967) have reported the occurrence of two modal groups in the width of the first growth band in scallops from Southern Ireland and Northern Ireland respectively, which they connected with two spawning peaks, autumn spawned individuals predominating in both areas. In Holyhead, North Wales, Baird (1966) found two modal groups but the spring spawned type predominated, though in one year, 1956, a failure of the early settlement resulted in a scarcity of scallops with large first growth bands.

It has not so far proved possible to confirm the origins of the two modal groups by direct study of settlement and the subsequent growth of spat. Despite the fact that both the spring and autumn spawnings of Manx scallops have been shown to result in the successful production of planktonic larvae, experiments in 1976 to study the seasonal settlement of spat failed to yield any that could be attributed to an autumn spawning (Brand, Paul and Hoogesteger, 1980). The authors imply that this might be due to differences in settlement behaviour between larvae from the two main spawnings such as are found in *Mytilus edulis* in northwest Spain (Andreu, 1968).

I found that in Loch Torridon (northwest Scotland), the majority (84%) of adult scallops had wide first growth bands,

28–49mm. Young spat 3–9mm long found there in October 1968 1968 grew slowly throughout the winter and did not lay down an obvious growth ring in the spring of 1969. These would almost certainly have laid down their first obvious ring in their second spring. This might lead to confusion, where late settlement occurs, between late settled animals and spring spawned scallops of the next year class (Mason, 1969).

In areas where scallops have prolonged spawning but no distinct peaks the first growth band would be expected to show a wide range of values but without a bimodal distribution.

The annual growth rate Observations on the size of animals at different stages of development may be used to derive a relationship describing the average growth of animals with time. The size Y_t of an animal at age t may by expressed in mathematical terms by a formula or function involving t and one or more parameters $\theta_1, \theta_2, \ldots\ldots$ These parameters reflect particular characteristics of the growth curve and different growth curves may be conveniently compared in terms of differences between the parameter values.

The choice of an appropriate function is, to some extent, arbitrary. Three-parameter functions are often used to describe biological growth, among the most frequently employed being the von Bertalanffy:

$$Y_t = Y_\infty (1 - be^{-kt})$$

and the Gompertz:

$$Y_t = Y_\infty \exp(-be^{-kt}).$$

The von Bertalanffy curve increases at a steadily decreasing rate and has no point of inflexion. It is based on the assumption that metabolic requirements increase as a function of the volume, and not a linear dimension, of the organism, and therefore predicts a slowing down of growth as the organism grows bigger. The Gompertz curve has a point of inflexion and the curve is asymmetrical about this point, which occurs where $Y = 0.37Y_\infty$.

The von Bertalanffy curve has proved suitable for describing the growth in length of many marine organisms, particularly over the later stages, while, perhaps because of the greater flexibility of its shape for smaller values of t, the Gompertz curve may, in some instances, provide a more satisfactory description of the earlier stages of growth. Of the three parameters involved in these curves k, the growth constant, is related to the rate of growth of the animal and Y_∞ is its final size. The constant b has no biological significance. In the von Bertalanffy curve the value of t corresponding to $Y_t = 0$ is $t_o = \log_e b/k$, the point where the curve cuts the time axis, but the Gompertz curve does not touch the time axis at any finite value of t.

There is a high degree of correlation between the length, breadth and thickness of the shell of *P maximus* (Mason, 1957). The coefficients of correlation were worked out on 614 Manx scallops of all ages from 0+ to 13+. The coefficient of correlation between length and breadth was 0.9937, and that between length and thickness 0.9597. Furthermore, a high degree of correlation exists between the length of the scallop and the breadth of the flat valve. The coefficient of correlation, worked out on 414 scallops of all ages from 0+ to 13+, is 0.9954. The scallop, in fact, grows proportionately in all dimensions, and retains virtually the same shape throughout its life, with the exception of a concavity on the upper valve during the first year or so. The annual increment of any one of these dimensions will, therefore, give a reliable indication of the rate of growth from one year to another.

Growth of Bradda and Bay Fine scallops was plotted and von Bertalanffy and Gompertz curves were fitted (Pope and Mason, 1980), using the following measurements: (*1*) position of successive annual growth rings on the flat valve, and (*2*) annual increase in length of the whole shell.

(*1*) Since each growth ring represents the position of the edge of the shell at the end of an annual growth period, it is possible to measure directly on the shell of any scallop the breadth of the flat valve of that scallop at the end of each growth period in its life. By measuring the distances of the various growth rings from the umbo at right angles to the hinge line it is possible to draw up a growth curve. The results are shown in *Table 6* and *Fig 22*. Only the first eleven rings were measured in autumn spawned scallops because of the crowding together of the later rings and the small numbers of older scallops obtained. Few spring spawned scallops were found with more than seven rings.

(*2*) Growth curves were drawn by plotting the length of scallops, measured during the cessation of growth in the winters of 1950–1951 and 1951–1952; figures for the two winters were combined. No scallop was measured which showed new growth

Table 6 Mean distances of the growth rings from the umbo (flat valve) of Port Erin scallops

Growth rings	*Mean distance (mm)*	
	Autumn spawned	*Spring spawned*
1	19.0	36.2
2	48.0	65.8
3	76.4	88.1
4	94.6	101.8
5	104.9	109.1
6	112.3	113.4
7	114.2	114.6
8	119.2	–
9	121.6	–
10	123.9	–
11	126.1	–

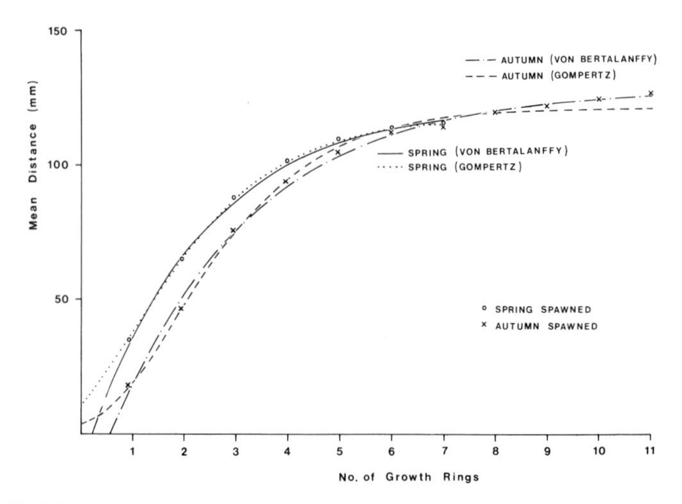

Fig 22 Mean distances (B) of growth rings (flat valve) of Port Erin scallops from the umbo and fitted von Bertalanffy and Gompertz annual growth curves. (*Pope and Mason, 1980*)

Parameters of fitted curves:-

	B_∞	k	t_o	b
Spring spawned				
von Bertalanffy	121mm	0.476	0.27	–
Gompertz	116mm	0.719	–	2.394
Autumn spawned				
von Bertalanffy	128mm	0.380	0.62	–
Gompertz	122mm	0.677	–	3.648

Functional forms:-

von Bertalanffy $Y_t = B_\infty\ (1 - e^{-k(t-t_o)})$

Gompertz $Y_t = B_\infty\ \exp\,(-be^{-kt})$

at the edge of the shell. In this method, a scallop which has no growth rings has completed its first growth period, one with one ring has completed two growth periods, and so on. Autumn spawned scallops with eleven or fewer growth rings and spring spawned scallops with seven or fewer rings were used. The results are given in *Table 7* and *Fig 23*.

Table 7 Length of Port Erin scallops at the end of successive annual growth periods

No. of completed growth bands	*Mean length (mm)* Autumn spawned	*Spring spawned*
1	21.2	37.5
2	53.5	73.3
3	87.7	98.0
4	108.1	114.7
5	118.6	118.9
6	128.0	134.4
7	131.8	134.2
8	136.8	140.8
9	137.9	–
10	142.8	–
11	142.1	–
12	148.1	–

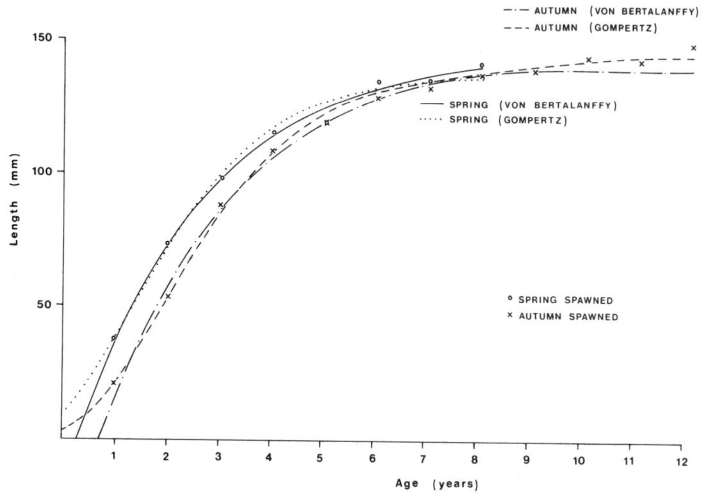

Fig 23 Length (L) of Port Erin scallops at the end of successive annual growth periods and fitted growth curves. (*Pope and Mason, 1980*)

Parameters of fitted curves:-

	L_∞	k	t_o	b
Spring spawned				
von Bertalanffy	146mm	0.396	0.24	–
Gompertz	138mm	0.661	–	2.512
Autumn spawned				
von Bertalanffy	147mm	0.372	0.62	–
Gompertz	141mm	0.658	–	3.651

Functional forms:-

von Bertalanffy $Y_t = L_\infty (1 - e^{-k(t-t_o)})$

Gompertz $Y_t = L_\infty \exp(-be^{-kt})$

The second method has several disadvantages. The dimensions at the end of a particular growth period can be obtained only by measuring scallops of that particular age; it is dependent on small-mesh dredge samples for younger scallops, and even then few are caught (see later); and it must use scallops caught during the cessation of growth. In the first method, on the other hand, all the growth rings can be measured on every shell; scallops from the commercial dredge can be used, since information about the early years of the scallop's life can be obtained from older shells; and scallops caught at any time of the year can be used. However, when commercial samples are measured on board boats, or on landing, it is usually possible from the point of view both of time and of convenience to measure only overall length. This is the most logical dimension to use should any size legislation become necessary, being the largest dimension and much the easiest for enforcement officers and fishermen to measure by means of a simple gauge.

Where it is possible to obtain measurements of rings, as in *Table 6,* it is possible to convert these into overall lengths by means of a simple linear equation. In Port Erin scallops this was found to be $L_t = 1.16B_t - 0.93$ where L_t = overall length and B_t = breadth of flat valve at age t.

In the first method, data from scallops of different year classes are grouped together, thereby masking any variation in growth rate there may be from year to year. Such variation could be shown by measuring one particular growth band, say the third, in scallops of all ages in one particular season. The second method would also afford a means of comparing the rates of growth in different years if sufficient scallops were measured to allow of keeping the figures for various winters separate.

Both the von Bertalanffy and Gompertz equations gave a reasonable fit. Von Bertalanffy was undoubtedly the better for the older scallops, which is the more important part of the curve from the point of view of population dynamics. The fitted von Bertalanffy curves agreed well with the observed values of length (L) and breadth (B) of older scallops and the calculated value of maximum length (L_∞) and breadth (B_∞). In the fitted Gompertz curve, however, calculated values of L_∞ and B_∞ were well below actual observed values.

At the lower end of the curves, too, the von Bertalanffy

equation gave a very good fit, the intercept (t_o) on the time axis in autumn spawned scallops being consistently larger than that in spring spawned scallops, reflecting the fact that the spring spawned scallops grew for almost a complete growing period and lived for almost a year before laying down the first growth ring, whereas autumn spawned scallops grew for a correspondingly shorter time before doing so. The Gompertz curve did not reflect this because it does not meet the time axis.

The growth curves show that, in autumn spawned scallops, growth is greatest in the second and third annual growth periods, and is approximately equal in extent in these periods. Thereafter the annual growth decreases progressively. Spring spawned scallops, which are larger in all dimensions at the end of the first growth period than autumn spawned scallops, grow most in the first two periods, in each of which growth is comparable with that of autumn spawned scallops in the second or third period. The annual increment decreases steadily in each growth period after the second. The mean dimensions of spring and autumn spawned scallops approach each other more closely with increasing age.

The growth curves obtained by the two methods are in good agreement, and are of a type characteristic of lamellibranch shells.

Comparison of the rates of growth of scallops from different areas and depths Differences in rate of growth between localities are frequently found in lamellibranchs. Growth curves have been derived for scallops from other areas, *eg* The Clyde Sea Area and West of Kintyre (Mason and Drinkwater, 1969), Southern Ireland (Gibson, 1956) and the English Channel (Franklin *et al,* 1980), and are shown in *Fig 24*.

Several factors have been suggested to account for local differences of growth rate in pectinid and other molluscan species, including currents, the nature of the sediment, temperature, the nature of the sea bed and food availability. Gibson (1956) found that scallops living on sheltered beds grew more quickly than those on exposed beds, and suggested that this is due to excessive particle bombardment interfering with feeding on the latter beds.

On comparing the growth rates of scallops in the five places off the southwest of the Isle of Man I found that scallops living in the shallowest water grew more quickly than those in deeper water

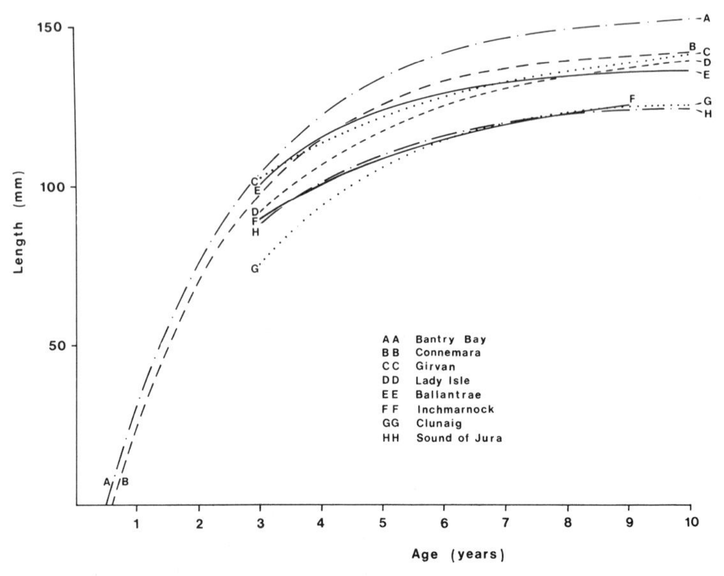

Fig 24 Growth curves of scallops from different areas

(*Fig 25*). It is suggested later that the growth of *P maximus* is influenced by temperature. It is of interest in this connection that from May to August, when the scallop is growing most rapidly, bottom temperatures inshore are higher than those offshore, resulting in a higher rate of growth on the shallower, inshore beds. Food may also be an important factor, phytoplankton being more abundant there.

Gruffydd (1974a) suggested that density influences growth, this being presumably due to competition for available food. Some interesting light is shed on this suggestion by a comparison of the growth of scallops in 100m depth off the Cock of Arran in 1974–1976 with that in 1965–1966. In the early 1960s scallops on this bed were literally smothered by queens, which in places lay three or four deep on the sea bed. By the early 1970s queens were scarce and many dead shells were present. The growth rate of scallops was greater after the reduction in number of queens (*Fig 26*), suggesting that their growth in the early 1960s might have been retarded by the queens competing for food or space.

The annual period of growth The annual period of growth of scallops was determined by measuring throughout the year the

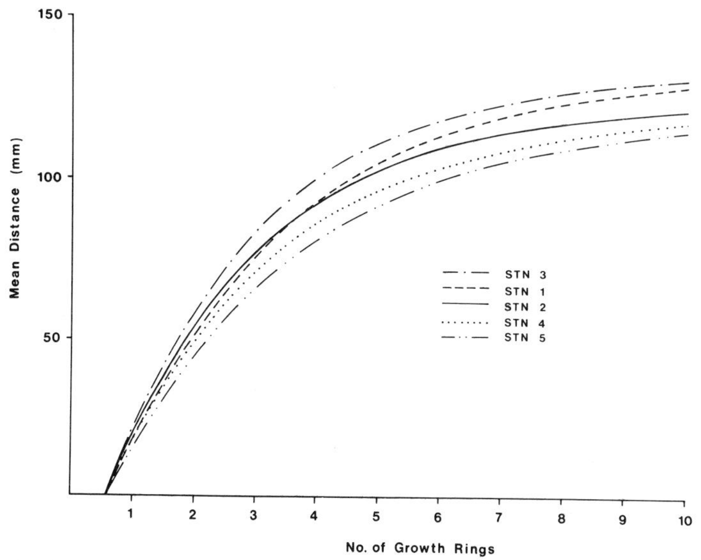

Fig 25 Growth of scallops at five stations off the south end of the Isle of Man. (*Fig 18*) (*Mason, 1957*)

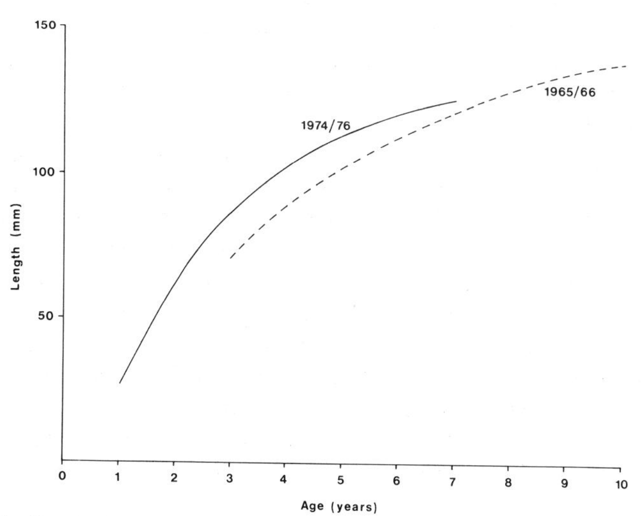

Fig 26 Growth of scallops off the Cock of Arran (Clyde Sea Area) during and after the great abundance of queens

width of shell outside the outermost growth ring on the flat valve of the shell of Bradda scallops, spring and autumn spawned scallops being treated separately. Autumn spawned scallops with 0–5 growth rings were used, but insufficient spring spawned scallops were obtained with more than three rings. Since the annual growth ring is laid down some time in March, April or May, it is necessary to consider the period from March of one year to May of the next year to cover a complete growth period.

The results (*Table 8* and *Fig 27*) show that growth commences in March, April or May, and is most rapid from June to September or October, when it begins to slow down, stopping altogether from December until the following March, April or May. These results agree well with those of Gibson (1956), who found that

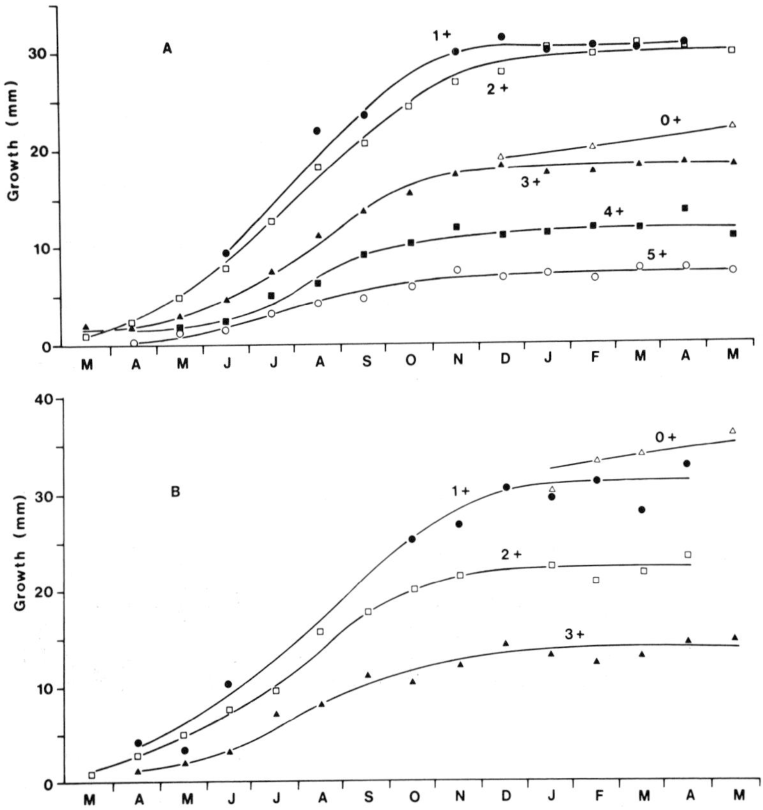

Fig 27 Growth of shell (mm) of *Pecten maximus* of different ages beyond the last-formed annual ring, throughout the year. (A) Autumn spawned and (B) spring spawned scallops. (*Mason, 1957*)

Table 8 Growth of shell (in mm) beyond the last-formed annual ring, throughout the year (Mason, 1957)

Number of growth rings		*M*	*A*	*M*	*J*	*J*	*A*	*S*	*O*	*N*	*D*	*J*	*F*	*M*	*A*	*M*
	Autumn spawned scallops, age-groups 0 to 5															
0	*Number measured*	–	–	–	–	–	–	–	–	–	1	1	2	–	–	4
	Mean growth	–	–	–	–	–	–	–	–	–	19.5	15.5	20.5	–	–	22.3
1	*Number measured*	–	–	–	4	–	2	3	3	9	9	6	12	31	7	–
	Mean growth	–	–	–	9.5	–	22.5	24.0	26.0	30.4	31.9	30.7	31.2	30.1	31.3	–
2	*Number measured*	1	27	38	20	15	8	9	7	35	5	9	50	5	37	12
	Mean growth	1.0	2.2	4.9	7.9	12.7	18.8	21.3	25.0	27.5	28.4	30.1	30.4	30.4	30.9	30.0
3	*Number measured*	3	24	74	127	36	17	80	66	52	19	45	90	49	45	9
	Mean growth	2.0	1.9	2.9	4.6	7.6	11.5	14.1	15.9	17.8	18.8	17.8	18.0	18.6	18.9	18.5
4	*Number measured*	–	29	49	52	16	8	13	24	9	8	16	12	13	5	22
	Mean growth	–	2.0	1.7	2.3	5.2	6.3	9.4	10.4	12.2	11.4	11.6	12.2	12.0	13.6	10.7
5	*Number measured*	–	3	12	58	43	27	22	20	32	9	35	22	21	33	16
	Mean growth	–	0.3	1.5	1.6	3.3	4.4	5.0	6.2	7.5	7.1	7.3	6.8	7.8	7.8	7.3
	Spring spawned scallops, age-groups 0 to 3															
0	*Number measured*	–	–	–	–	–	–	–	–	–	–	1	2	13	–	2
	Mean growth	–	–	–	–	–	–	–	–	–	–	30.0	33.5	34.2	–	36.5
1	*Number measured*	–	1	1	1	–	–	–	5	7	8	2	13	8	5	–
	Mean growth	–	4.0	3.0	10.0	–	–	–	25.2	26.6	30.4	29.5	31.2	28.1	33.0	–
2	*Number measured*	1	19	10	6	5	12	7	5	2	–	11	6	32	13	–
	Mean growth	1.0	2.5	4.7	7.3	9.4	15.8	17.6	19.8	21.5	–	22.5	20.8	21.7	23.7	–
3	*Number measured*	–	1	13	10	6	1	12	17	15	6	8	9	7	8	2
	Mean growth	–	1.0	1.8	3.0	6.8	8.0	10.7	10.3	12.1	14.0	13.1	12.1	12.9	14.5	15.0

growth of *P maximus* in Irish waters ceases from November to February.

With two exceptions, all the scallops which had not yet deposited their first growth ring were obtained between December and May, after the completion of their early growth. The exceptions were two scallops, 3.5 and 3.0mm long, which were found on the Bradda bed in August, 1952. These probably arose from the small spawning of July, 1952, but since this is not certain they were not included in *Table 8* and *Fig 27.* Later work in northwest Scotland (Mason, 1969) suggested that growth might continue slowly throughout the first winter of the animal's life.

Age and growth in *Chlamys opercularis*

As in *Pecten maximus,* the shell of *Chlamys opercularis* shows striae on the outer surface as a result of the method of deposition of the shell (Broom and Mason, 1978). At regular intervals these are closely packed together to form distinct rings, though less distinct than in *Pecten maximus* (Taylor and Venn, 1978), which become apparent in the spring when growth is resumed after a winter cessation or accelerates after a winter slowing down. This has been noted in a wide variety of localities, including the English Channel (Pickett and Franklin, 1975; Broom and Mason, 1978), the Isle of Man (Soemodihardjo, 1974) and the Clyde Sea Area (Taylor and Venn, 1978). Aravindakshan (1955) postulated that the first two rings were caused by spawning, but this has been discounted by subsequent workers. However, Pickett and Franklin (1975) and Taylor and Venn (1978) said that an extra disturbance ring might be laid down by English Channel and Clyde queens respectively at the time of spawning, though it was less distinct and easily distinguished from the true winter ring. Similarly, Broom and Mason (1978) found that a disturbance ring was deposited in English Channel queens held in cages or in tanks each time they were removed for measurement.

As in *P maximus,* the first ring shows a wide range of positions, and in some areas, *eg* Port Erin, Isle of Man (Soemodihardjo, 1974) and the Clyde (Taylor and Venn, 1978), a bimodal distribution which has been attributed to two mass spawnings or settlements per year. The total range in Manx queens was 12–40mm and the peaks of ring position at Port Erin were at 18–22mm (the larger peak) and 30–34mm, suggesting that animals with the smaller first growth band resulted from a major autumn spawning and those with the larger first band from a minor summer spawning (Soemodihardjo, 1974). The range of breadth of the first growth band in Clyde queens was 12–42mm. Subsequent differential growth results in the bimodality being lost in older queens (Taylor and Venn, 1978).

There is some evidence that during the queen's first one or two winters growth does not stop completely but continues at a much reduced rate, accelerating again in the spring. It has been suggested that spat which settles late in the year may be so small that growth may not slow down sufficiently to lay down a ring, or it may be so indistinct as to be confused with other marks. This could mean that the first distinct ring is laid down in the second spring and could result in such late autumn spawned queens being confused with spring spawned queens of the next year class (Soemodihardjo, 1974; Paul, 1978). Indeed Soemodihardjo suggested that if the first band in Manx scallops was more than 30mm wide the first ring was likely to have been laid down when the queen was more than 1 year old.

Chlamys opercularis is relatively short-lived. While Taylor and Venn (1978) found some individuals with up to eight annual growth rings, few with more than four or five rings were found anywhere. Similarly, in most areas queens do not exceed 80–85mm, though the northern isles Shetland (Mason, 1980) and Orkney are exceptions, and at Orkney specimens up to 105–109mm long are recorded (Mason, 1970).

Most workers have found that the von Bertalanffy growth equation fits the growth data well and have used it to describe the linear growth of the queen (Soemodihardjo, 1974; Broom, 1976; Broom and Mason, 1978; Taylor and Venn, 1978; Mason, Shanks, Fraser and Shelton, 1979; Mason, 1980), though Taylor and Venn found that the Gompertz equation also provided a good description.

Taylor and Venn (1978) tabulated the growth rate of queens from various areas of the British Isles and noted the similarities between Manx and Clyde queens, where a commercial size of 55mm (breadth, though length is only slightly different) is reached by the end of the third growth period. Faster growth rates were found off the east coast of Eire (Kish Bank–Bhatnagar, 1972), off Yorkshire (Pickett and Franklin, 1975) and at Shetland (Mason,

1980). Growth in the Plymouth area of southwest England is considered poor by Pickett and Franklin. Everywhere growth is most rapid during the first two or three growth periods and subsequently slight (*Fig 28*).

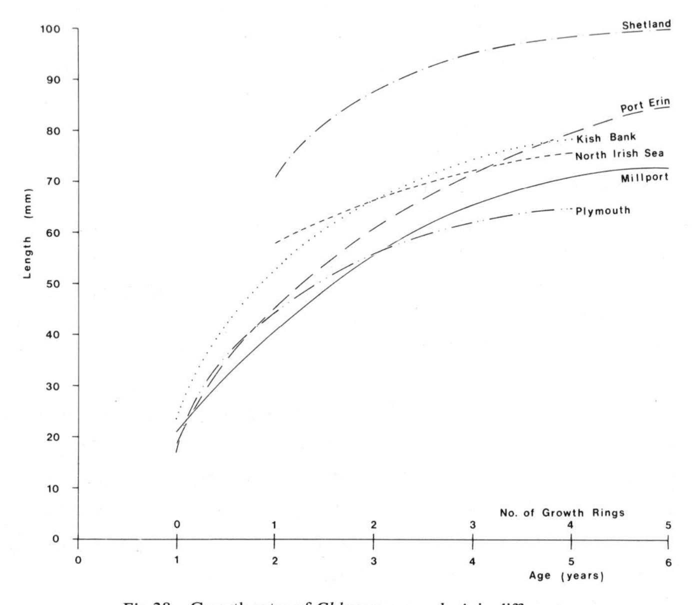

Fig 28 Growth rates of *Chlamys opercularis* in different areas

As in *Pecten maximus,* growth in *C opercularis* slows down or stops completely in the winter, recommences in the spring, when the annual growth ring becomes obvious, is rapid in the summer and slows down in the autumn to the winter minimum or cessation (Broom and Mason, 1978; Taylor and Venn, 1978; Pickett and Franklin, 1975).

Broom and Mason (1978) measured the size of individual queens at intervals over a period of 18 months and counted and measured the striae laid down between the times of measurement. Early in the year few striae were laid down relative to the number of days and they were close together (0.11mm). As midsummer approached the number of striae increased and reached 80–90% of the number of days, but never exceeded the number of days, and then fell away again. At the same time the distance apart of striae increased to a maximum of 0.26mm in May–June and fell away to October, when there was a sudden short increase corresponding to the autumn phytoplankton peak followed by a decline in winter. Broom and Mason suggested that lamellae are probably laid down daily, as in some other pectinids (Clark, 1968), whenever conditions are favourable. There is recent evidence that in French waters the striae of *Pecten maximus,* at least during the first two years of life, are laid down daily (Antoine, 1978), though Gruffydd (1981), on the basis of experimental studies, found no evidence of a daily rhythm.

Causes of cessation of growth

The growth of a marine bivalve is the result of many interacting factors, of which temperature and food are well known to be important. It has been suggested that the annual cessation of growth which results in the formation of the growth ring is related in some pectinids to poor condition after the formation of gametes and spawning.

Spawning cannot be an effective factor in causing the winter cessation of growth in *Pecten maximus* off the Isle of Man since the times of spawning occur during the period of growth, while during the winter cessation of growth there is no spawning (Mason, 1958a). Indeed *P maximus* does not spawn until after the deposition of its second growth ring (Mason, 1957).

Mason (1957) showed a close relationship between linear growth of *Pecten maximus* and both temperature and rate of feeding, as shown by solid matter in the stomach (*Fig 29*). Broom and Mason (1978) found that temperature and food play a part in making conditions right for growth in *Chlamys opercularis*. They found that most lamellae were deposited and the striae were further apart, *ie* most linear growth occurred, about midsummer and again in the autumn. They found that while significant correlations were shown between growth and temperature and growth and food availability (as shown by the abundance of chlorophyll in the sea water) the most highly significant correlation was between growth and the product of temperature and food.

The dry weight of the adductor muscle in both *Pecten maximus* (Comely, 1974; Connor, 1978) and *Chlamys opercularis* (Soemodihardjo, 1974; Taylor and Venn, 1979) is lowest in the spring when water content is highest, increases in the summer to a maximum in the autumn, and falls through the winter to the following

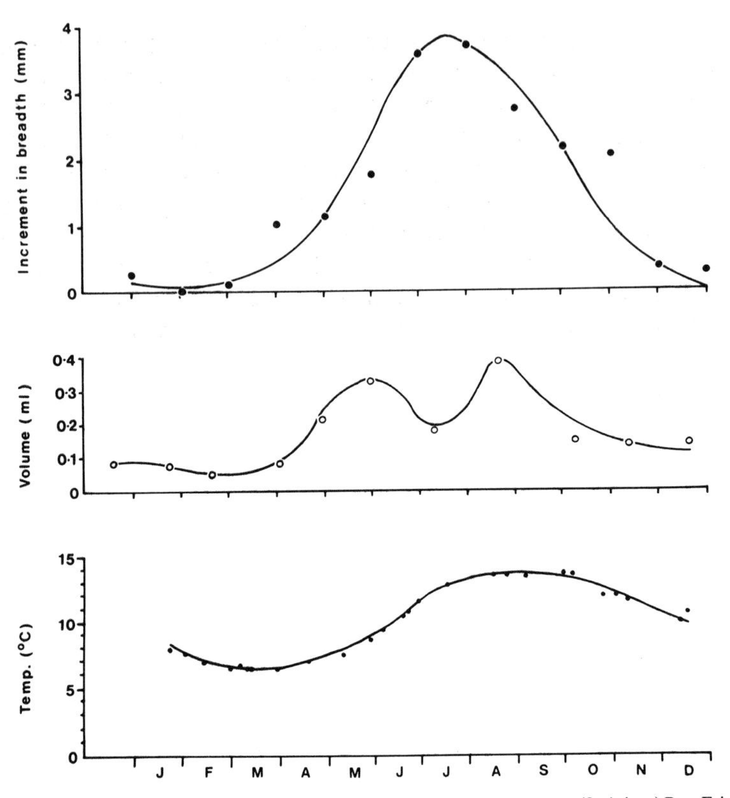

Fig 29 Mean monthly increment in breadth of flat valve of *Pecten maximus* (2–4 rings) Port Erin, volume of solid matter in the stomachs of 1948 brood scallops in 1952, and bottom temperatures on the scallop beds in 1951. (*Mason, 1957*)

spring. The changes are due to changes in carbohydrate and protein content. A similar cycle is found in tissues other than the gonad, but carbohydrate is low throughout the year and changes are due to changes in protein and lipid content. The dry weight of the gonad is least after the spawning, seasonal variations being due to the development and eventual release of gametes, gametogenesis recurring in the winter when food is scarce and being supported by reserves of glycogen and protein in the adductor muscle which are not, however, sufficient to allow growth.

The growth of poikilotherms tends to increase with temperature because of the increase in metabolic rate, provided enough food is available. Thus in *P maximus* and *C opercularis* linear growth recommences in the spring and is maintained throughout the summer and slows down in the autumn as temperature falls. The energy maintenance requirement depends, however, not only on temperature but also on size. The von Bertalanffy growth equation predicts a slowing of growth rate as the animal increases in size, based on the assumption that metabolic requirements increase as a function of the volume. With increasing size it is more and more difficult to gather food in excess of increased maintenance needs. This might partially explain why queens may grow, albeit slowly, over the first winter or two while in subsequent winters growth virtually ceases (Broom and Mason, 1978).

9
The state of the fisheries and stocks

As was indicated in *Chapter 2,* there has been a large expansion of scallop fishing around the British Isles during the past 20 years, and this has been augmented by a completely new fishery for the queen. As more and bigger boats have entered the fisheries, more concentrations of scallops have been found. In the English Channel boats are fishing further offshore. Perhaps surprisingly in view of the length and intensity of fishing, new stocks are still being found off Scotland, the latest major discovery being the bed on Smith Bank in the Moray Firth, exploitation of which commenced in 1978 and is still proceeding. Similarly, new queen grounds have been discovered as boats have ventured further afield. There continues to be a ready market for both species, despite competition from the cultivated Japanese scallop (*Patinopecten yessoensis*).

There might appear to be a danger, especially with species such as the scallop and the queen, both of which compared with fin fish are relatively sedentary and confined to a particular type of sea bed, that the beds might become overfished. Indeed, catch rates on newly exploited beds have inevitably fallen from their initial high levels, and this has given rise to complaints of overfishing and demands for legislation to protect the stocks. In order to assess the state of stocks it is desirable to have runs of reliable catch composition and catch per unit effort (cpue) data extending over many years. The fact that such runs are not available for a particular stock should not be used as an excuse for inaction if all the signs are that a stock is in danger, but it should be emphasized that many factors need to be considered. Good series of data are now becoming available for the longer-standing Scottish scallop and queen fisheries, and preliminary stock assessments have been carried out on some of them (Mason, Nicholson and Shanks, 1979; Mason, Shanks and Fraser, 1980, 1981; Mason, Shanks,

Fraser and Shelton, 1979) and certain conservation measures recommended. The history of the main Scottish fisheries since the mid 1960s reveals some interesting trends.

Recent trends in the main Scottish scallop fisheries

Southwest Scotland

This region, comprising the Ayr and Campbeltown Fishery Districts, includes the Clyde Sea Area, and has been the major locus of the fishery since the 1930s. Landings reached a peak of 3,644 tonnes in 1969, declined to 584 tonnes in 1972 as queen fishing expanded, and subsequently recovered to a peak of 2,055 tonnes in 1976, since when they have declined little. Landings in the Clyde and west of Kintyre separately showed similar trends (*Fig 30*).

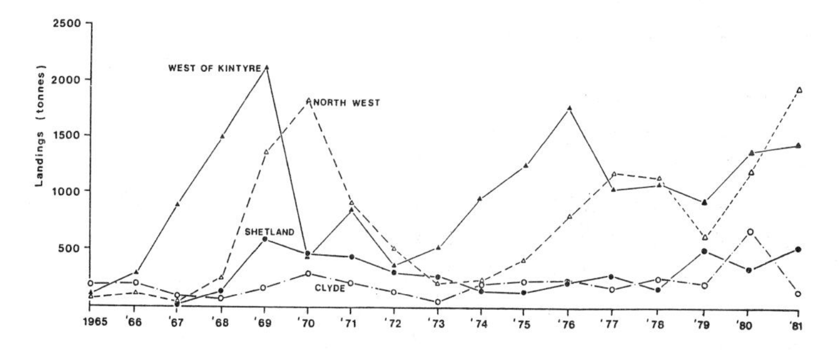

Fig 30 Scottish scallop landings by area, 1965–1981

Clyde Sea Area *Figure 31* shows the age composition of scallops caught by commercial dredge during successive winters. The data are for October to March, when little or no growth occurs. It is arbitrarily assumed that the new growth ring appears at the edge of the shell annually on 1st April when growth recommences.

In the late 1960s no dominant age group was evident in the catches from the east side of the Clyde. The high proportion of older scallops, especially those with more than nine rings, suggested that in a stock which had long been exploited there had been no successful recruitment for many years (Mason and Drinkwater, 1973). The appearance of more scallops with 4–6 rings in 1970–1971 and 1971–1972 indicated improved recruitment, and the more recent high proportions of 3–5 ring scallops indicate that improved recruitment has continued.

The 1961 age group had dominated the catches on the west side of the Clyde in the mid and late 1960s (Mason and Drinkwater, 1973) and was still discernible as the 7-ring group in 1968–1969 (*Fig 31*). Very few younger scallops were present in the catches during that period, suggesting that this particularly good brood was followed by a succession of poor ones. Again in 1970–

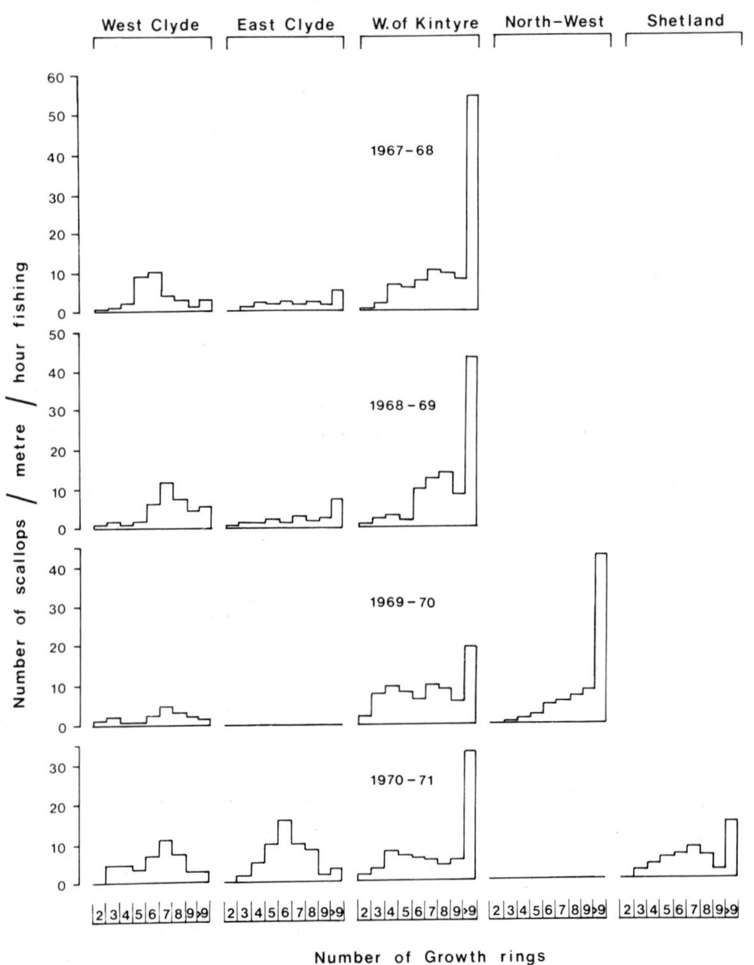

Fig 31 Scottish scallop catch; numbers of each age group taken per unit of effort, 1967–1981 (*and facing page*)

1971 and 1971–1972 there were signs of improved recruitment on the west side of the Clyde as young scallops appeared in the catches, and this has been maintained as 3–5 ring scallops have since been abundant.

Catch per unit effort data were obtained over a period of years from selected boats and are presented in *Fig 32.* In the late 1960s catch per unit effort was low on both sides of the Clyde, where the stock had been exploited for many years and recruitment had

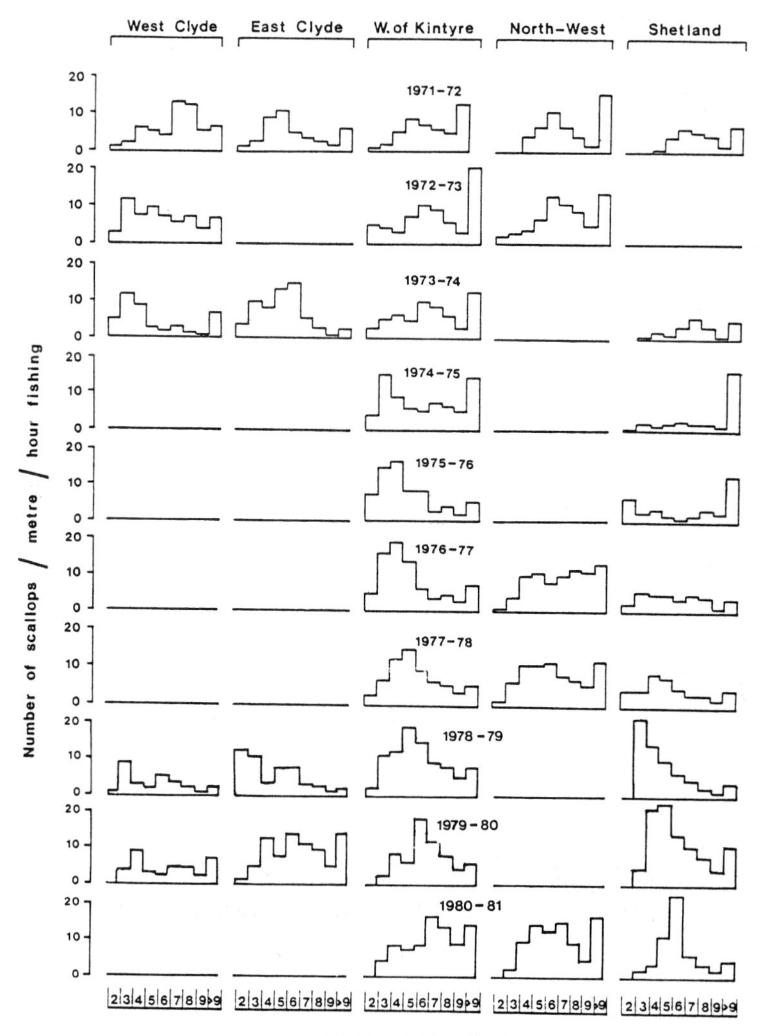

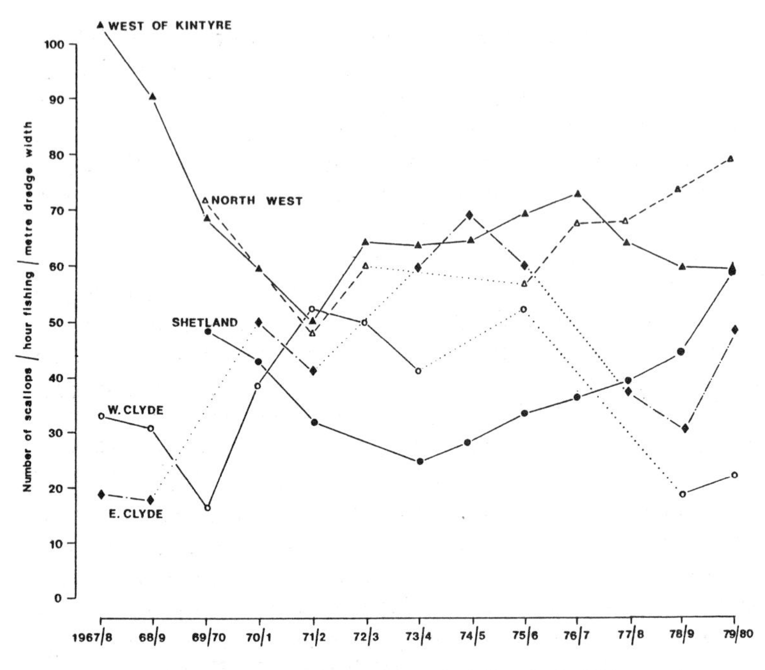

Fig 32 Scottish scallop catch per unit of effort, 1967–1980

been poor, and this led to fishermen expressing fears of overfishing. With increased recruitment in the 1970s, the catch per unit effort has also increased. At times the data for the Clyde have been limited, and in some years none were available. This is because fishing effort has recently been low in the Clyde, especially on the somewhat limited east side grounds, owing partly to more lucrative fishing elsewhere and partly to the newly developed queen fishery.

West of Kintyre This stock of scallops was virtually unfished until 1966–1967 but has since been heavily exploited. At first there was a high proportion of older scallops (seven or more rings) in the catches, and this was still apparent in 1968–1969 (*Fig 31*). As the fishery continued older scallops were removed and the age composition of the catch has come to resemble closely that in the Clyde. Recruitment appears to have been good in this area also in the 1970s, as shown by the recent high proportion of 2–5 ring scallops in catches.

Catch per unit effort was initially very high, being 103 scallops per hour per metre dredge width in 1967–1968. It declined rapidly, as would be expected in a newly-exploited stock, until an equilibrium was reached between fishing and recruitment. The lowest value, 50 per metre per hour, was reached in 1971–1972, but it has risen since on account of the recent good recruitment (*Fig 32*).

Northwest Scotland

Dredging, chiefly in the South of Mull and around Skye (*Fig 3*) first became appreciable in 1969, the biggest landings, 2,547t, occurring in 1970. Dredging decreased about this time, and despite the beginning of diving for scallops and its subsequent increase, landings declined to 284t, virtually all diver-caught, in 1974. They then increased as dredging recommenced, reaching 1,193t in 1977 and 1,142t, with a record value of £762,203, in 1978 (*Fig 30*). The catch from this hitherto unfished stock, like that west of Kintyre, at first contained a high proportion of old scallops, but here too the proportion has fallen (*Fig 31*). In the early 1970s, catch per unit effort, never very high, fell slightly; the recent presence of good numbers of 4- and 5-ring scallops in the catches suggests improved recruitment and catch per unit effort has increased (*Fig 32*).

Shetland

Dredging started in 1968, and after reaching 599t (value £74,000) in 1969 landings declined steadily to 96t (value £27,000) in 1973. They then recovered to 224t (value £114,000) in 1977. The age composition of catches has shown a similar trend to that in northwest Scotland, though there is still a fair proportion of older scallops (*Fig 31*). Catch per unit effort has generally been lower than elsewhere and declined during the early 1970s, though it has increased recently with signs of improved recruitment (*Fig 32*).

Conclusion

Thus similar trends have appeared in all the main Scottish scallop fisheries, namely, decline in total catch and catch per unit effort in the early 1970s owing to the effects of exploitation and a succession of poor broods, followed by increased recruitment and an associated increase in catch per unit effort in the mid and late 1970s. The conclusion must be that the stocks are at present in no danger of serious depletion.

A careful watch is, however, being kept on the Scottish scallop stocks, so that we might be informed should the situation change. On the heavily fished southwest Scottish grounds pre-recruit (1+ and 2+) scallops are monitored by means of a special sampling dredge with closer teeth spacing (50mm) and smaller belly ring and netting mesh (57mm) (Drinkwater, 1974) which shows up the occurrence of a good or bad year prior to recruitment (Mason and Drinkwater, 1974, 1975, 1976). There is some evidence from recent French research that the density of settlement of scallop spat on artificial collectors might be used to predict the strength of the recruitment to the fishable catch two to three years later (Buestel, Dao and Lemarié, 1979).

Yield assessments require some knowledge of the natural mortality (M) of the scallop. No direct measure is available for Scottish beds, but in assessments on some Scottish stocks a value of 0.15 has been used, which is in good agreement with values estimated for an unfished population in the North Irish Sea by Gruffydd (1974b). Another factor which has to be considered is mortality due to damage caused by dredges to scallops which are not actually caught (Gruffydd, 1972).

The studies (Mason, Nicholson, and Shanks, 1979; Mason, Shanks and Fraser, 1980, 1981) suggested that, while the Scottish stocks are currently in a healthy state, higher long-term yields per recruit could be achieved by an increase in the fishing effort and an appropriate change, usually a decrease, in the age at first capture. However, this would be achieved at the expense of a considerable decrease in breeding stock biomass per recruit with the consequent possibility of a recruitment failure and a consequent lower overall yield. In the Clyde particularly, the breeding biomass is already low. It was suggested that if it were considered necessary to protect the stocks this would be achieved on all the beds studied in Scotland by delaying age at first capture. It is suggested that this would be best done by introducing a minimum legal landing size of 110mm overall length, which in these areas effectively means not capturing scallops below four or five years old, and these would have spawned in at least three years before being exposed to capture. Some Scottish processors and fishermen have introduced a voluntary minimum of 100mm, which corresponds to an age of 3 to 4 years.

The meat yield is increasing most rapidly up to a size 110–120mm in Manx (Mason, 1959b; *Fig 33*) and southwest Scottish waters, and it would appear sensible to allow them to grow to something approaching this size before capture. This would fit in well with the requirements of the processors, there being most demand for scallops of 110mm and over. A Manx scallop 120mm long with a full gonad yields on average 32g of meat, of which the muscle accounts for 24g and the roe for 8g. Indeed, in the Isle of Man legislation first based on the trade requirements prescribes minimum landing size of 110mm. Delaying capture to this size will have a conservation effect in that it ensures that scallops spawn in at least two or three years before removal from the stocks. In areas of poor growth, such as the southwest of England, where few scallops reach 110mm, different considerations apply and smaller scallops must be landed, albeit at a lower price.

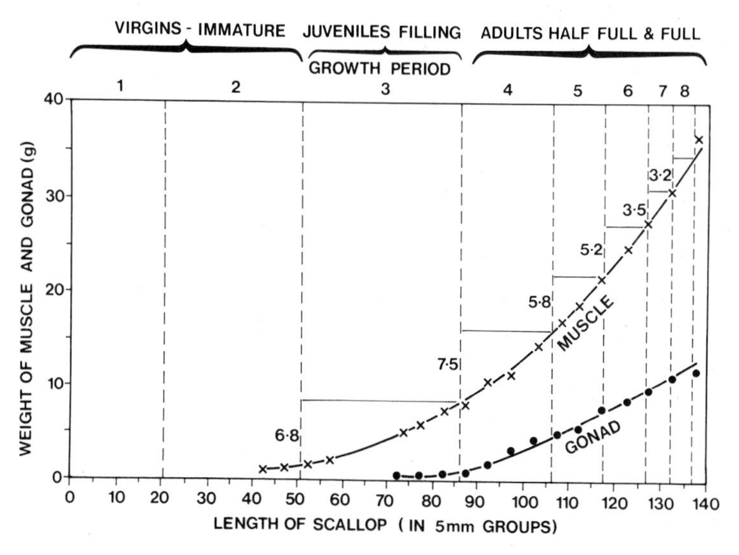

Fig 33 Increase in meat weight of the adductor muscle and gonad of Port Erin scallops with increasing length and age, and corresponding states of maturity of the gonad. Figures on the dotted lines show the actual increase in each growth period. (*Mason, 1959b*)

Recent trends in the main Scottish queen fisheries

Landings in the three main fisheries, North Irish Sea, Clyde and Shetland, have fluctuated owing to market conditions (*Fig 34*). Catch per unit effort has been well maintained, and even shown

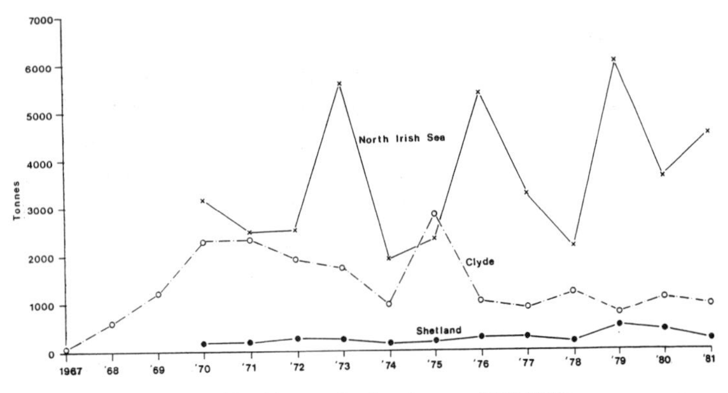

Fig 34 Scottish queen landings by area, 1967–1981

an upward tendency in the North Irish Sea (*Fig 35*). An assessment of the state of the stocks in the most heavily exploited Scottish fishery has suggested that the level of exploitation is not excessive from the points of view of growth and recruitment. Indeed it is possible that a moderate increase in fishing effort would lead to a greater long term yield while maintaining a satisfactory level of recruitment (Mason, Shanks, Fraser and Shelton, 1979). No detailed assessments have yet been made for the other main Scottish queen fishing areas, but the stocks appear to be in a reasonably healthy state as judged by recent landings and catch per unit effort data. With an animal such as the queen, with a rapid early growth rate and a short life span [and therefore a high natural mortality – values of from 0.7 in Scottish queens (Mason, Shanks, Fraser and Shelton, 1979) to between 1.5 and 2.0 in Channel

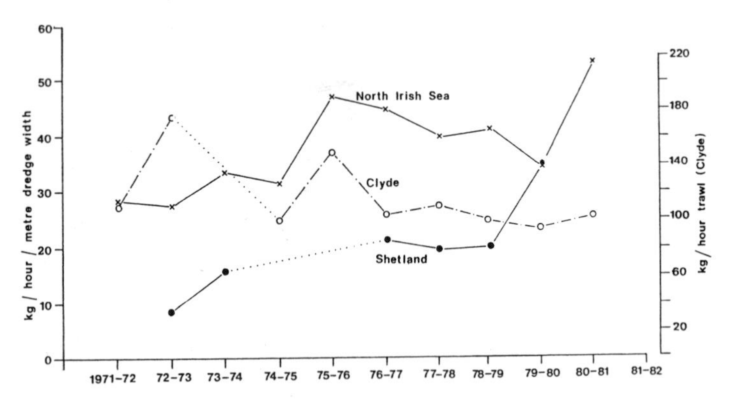

Fig 35 Scottish queen catch per unit effort, 1971–1982

Island stocks (Askew, Dunn and Reay, 1974) have been suggested], the best fishing strategy appears to be the one at present applied, namely to fish the stocks hard while they are there and before they die from other causes such as heavy predation by starfish. Judging by the consistently abundant settlement of queen spat obtained in a variety of places and the adult queen's abundance compared with that of the scallop, the queen is a very successful species and less subject to recruitment failures.

General considerations

Few details about the state of scallop and queen stocks in other parts of the British Isles are so far available, but there is no reason why the principles applied above to the Scottish stocks should not apply elsewhere.

There is, therefore, no evidence of serious danger to the stocks of either scallops or queens, even the long-exploited scallop beds of the Clyde Sea Area. The stocks, especially of queens, consist largely of discrete concentrations and experience has shown that when the catch rate of either species in one area declines to a level at which it is becoming uneconomical to fish, the fishermen move onto other grounds or even turn their attention to the other species or to other fish or shellfish, so allowing the grounds to recover. Recovery is especially quick in the case of the queen. Queen spat appears abundant in most years and growth is rapid in the first two to three years of life. In fact, the recovery of a bed has been demonstrated by the increase of catch per unit effort of Clyde stocks during the summer as the 2+ year group grow to commercial size and enter the fishery (Anon, 1978) (*Fig 36*).

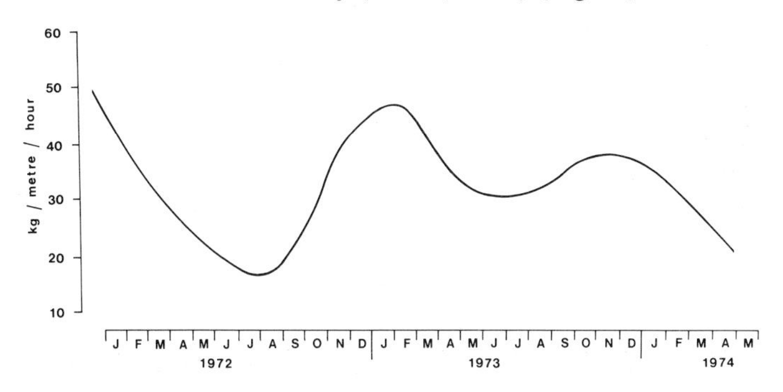

Fig 36 Seasonal catch per unit effort of queens in the Clyde Sea Area

10
Protective legislation and conservation

Few major fisheries today continue without some form of regulation. A fishery may be allowed to expand without restriction until there is unacceptable competition between the fishing units or until there are fears or evidence of over-exploitation. However, management measures may be introduced in anticipation of future problems. Many fisheries are controlled by one or more of a number of types of measures, *eg* mesh (or tooth-spacing) regulations, minimum legal size, closed areas or closed seasons and quotas. The choice is made often without any scientific basis, to serve practical or social needs, or as a consequence of intuitive reasoning. Many such regulations are retained for fear of the consequences of interfering with the equilibrium of a fishery (Hancock, 1979). All the regulations currently in operation in the scallop fisheries in the British Isles come into this category.

Not all the measures are suitable for scallops and queens, partly owing to the fact that the fisheries are fragmented, prosecuted in remote areas, and catches are landed in a large number of out-of-the-way places. Closed areas would be particularly hard to enforce. Since the distribution of both species is so patchy, adjacent grounds may vary in the degree of protection they require. Closed seasons are not necessarily effective since the stocks could be given a severe beating when they are open, and the total effort, as regulated at present by economic viability, might be very little reduced. Quotas or restrictions on total catch could be self defeating if, say, a fleet of boats discovered a bed of scallops with large numbers of young, recently-recruited animals which gave an easy and quick way to realize their weekly or monthly quota, so freeing them to engage in other types of fishing. Probably the best method of conserving scallop stocks, and that most commonly practised, is the minimum legal landing size, with or without the support of dredge tooth-spacing, mesh and ring regulations. The latter presupposes that scallops captured and returned to the sea survive, and recent work suggests that scallops, especially those taken in short dredge hauls, do have a high rate of survival (Chapman *et al*, 1977). Increasing the ring and mesh size and tooth-spacing results in fewer small (<70mm) scallops being caught (Drinkwater, 1974). Of the various types of regulation considered, the easiest to enforce would be a minimum legal landing size, checks being carried out at ports or processing factories.

In the Republic of Ireland scallop fishery (Gibson, 1956) there was formerly a closed season that restricted fishing to a period in winter and spring which varied from area to area to ensure that scallops were fished when the roes were in the best condition. Because of the variability in time when roes are full at different places and areas, which rendered the regulation of doubtful value and hard to enforce, the relevant bye-law was repealed. The only regulation now in force (Duff, 1976) is a minimum legal size of 4½ inches (114mm) 'measured across its greatest width', *ie* overall length. Skin diving for all shellfish is prohibited.

The only scallop regulation in Northern Ireland is a minimum legal size of 11cm measured 'in a straight line across its greatest length', *ie* overall length.

Scallop legislation in the Isle of Man has a complicated history. From 1943 a minimum legal size of 4½ inches (114mm) was introduced, together with a closed season which restricted fishing to to the period 1st October – 31st March with a possible extension to 30th April, and was aimed at preventing marketing in summer. As summer marketing became feasible, the close season fell into disuse. However, with the great increase in fishing effort in the early 1960s, and in the absence of statistics of landings, concern was felt for the stocks and it was decided that a series of *ad hoc* measures was justified. Furthermore there was pressure from merchants for a minimum landing size. In 1963, therefore, a series of bye-laws was promulgated. A closed season was reintroduced, allowing fishing only from 1st October to 30th May, a minimum legal size of 4½ inches 'across the broadest part of the *flat* shell' was introduced (this is less easy to measure than the overall length, and is slightly less than the overall length), and it was stipulated that spawned scallops with under-developed roes should not exceed 20% of the landings (surely a subjective judgement and one difficult to enforce). Subsequently the legislation was strengthened by including 'importing' and 'processing' as well as 'taking' and 'landing' scallops; the fishing season was extended by one day to 31st May and subsequently shortened by starting on 1st November; and the minimum size has been defined in metric terms as 11cm across the broadest part of the flat shell, which is roughly equivalent to the original overall length of 4½ inches.

There are no national regulations regarding scallop fisheries in England, Wales and Scotland. Fishermen in the southwest of Scotland for some time operated a voluntary minimum overall length of 4 inches (102mm), though this was partly in response to pressure from processors, who did not want smaller animals. The only local bye-law is that of the Devon Sea Fisheries Committee which prescribes a minimum size of 4 inches (102mm) overall length; scallop fishing is covered in some instances by limitations by local Sea Fisheries Committees in England and Wales on the size of boat permitted to fish in their waters (Franklin *et al,* 1980).

Should it become necessary to introduce protective measures for scallop stocks the regulations will have to be tailored to the individual stocks. For instance, a minimum legal size of 90mm overall length, which has been suggested as desirable to protect the Plymouth stocks, would have no beneficial effect on the southwest Scottish stocks, where a 110mm overall length was suggested to safeguard recruitment. Conversely, a minimum size of 110mm in the Plymouth fishery would mean that very few scallops could be landed. A problem of enforcement would be posed by the fact that boats from one part of the British Isles may fish in other parts and return home to land their catch.

To avoid confusion, the same dimension should be used throughout for minimum size regulations. The most logical, since it is easiest to measure and least prone to error, is the overall length of the animal, which is equivalent to the length of the round valve.

It is so far considered that, at the present levels of exploitation, British queen stocks need no protection and that the best fishing strategy, because of their rapid early growth and high natural mortality, is to fish them hard and catch them before some other predator gets them. Sometimes however, processors have specified that they will not take queens below a certain size, usually 45 or 50mm (or even 55mm) owing to the extra work involved in handling more queens in order to achieve a given meat yield.

11
Cultivation and stocking

Man's use of the production of the sea has hitherto largely been confined to the exploitation of natural production. However, in addition to consideration of methods of fishery management, attention is now being turned increasingly to cultivation or farming as a means of increasing production. Whereas fisheries are wholly dependent on fluctuating natural resources, cultivation has the advantage of increased stability of supply and price, which increases economic efficiency and attracts capital investment.

Unlike many of the fin fish species exploited by man, filter-feeding bivalves are low in the food web, feeding directly on the primary producers, phytoplankton and bacteria, as well as on non-living organic matter. For every stage an animal is removed from the basic production of carbon there is an energy loss of 80–90% in converting food into flesh. Therefore, while predatory fish caught by men are relatively wasteful of the basic production, filter-feeding bivalves are more efficient converters and so lend themselves well to cultivation, producing greater yields per unit area than species at higher trophic levels. Their sessile or semi-sessile way of life enhances their suitability. They do not waste energy pursuing food, which is brought to them by currents. They filter food from a great volume of water, with the result that small areas of sea can be made to yield large quantities of bivalve flesh. Furthermore they are easier to harvest than fin fish and frequently require less capital investment for equipment, buildings and space than motile forms (Mason, 1976).

Iversen (1968) defined sea farming as 'a means to promote or improve growth, and hence production, of marine and brackish-water plants and animals for commercial use by protection and nurture on areas leased or owned.' Kinne (1970) distinguished four ascending classes of cultivation, each of which includes those preceding it:–

(*1*) maintenance: keeping organisms alive, without significant growth, for scientific or commercial purposes;

(*2*) raising: bringing up (fattening) young adults;

(*3*) rearing: bringing up early ontogenetic stages (*eg* fertilized eggs, larvae);

(*4*) breeding: production of and bringing up offspring.

The switch from free fishing to organized cultivation is radical, involving the lease or ownership of an area to which the operator has sole rights; private rights are essential to farming.

Settlement of bivalve larvae and young post-larvae is commonly promoted by presenting a suitable substratum. These young may be transferred to areas where food is abundant and growth consequently fast, and they are marketed at an appropriate size. Growth is regulated by controlling density in relation to food supply. Where necessary and practicable, the cultivated species is protected from enemies and competitors. This method, which is commonly used for cultivating mussels (Mason, 1972a, 1976), falls into Kinne's second class.

Techniques of hatchery breeding and the rearing of larvae, which fall into Kinne's fourth class, have been developed successfully for some bivalves, notably oysters. Such techniques are, however, often difficult to develop and establish and costly to maintain, so that hatchery rearing is economically feasible only for high value species such as oysters (adult natives, *Ostrea edulis,* fetch £0.35–£0.40 each at first sale). Hatchery-based cultivation of scallops (*Pecten maximus*), which in 1980 fetched £3 per dozen for good-sized ($\geqslant$ 120mm) individuals, might be economically feasible, but hatchery-based cultivation of queens (*Chlamys opercularis*), which fetch only about 1.5 or 2p each on average, almost certainly would not. Queen cultivation, like that of other relatively low-value species such as mussels (Mason, 1972b, 1976), would have to be based on the collection of naturally occuring seed.

Experimental techniques for hatchery rearing larvae of *P maximus* from the egg to metamorphosis have been developed (Comely, 1972; Gruffydd and Beaumont, 1972; Buestel *et al,* 1978; Sasaki, 1979). Survival was good initially (Sasaki, 1979), some 10% reaching settlement size in three weeks at 18 °C, though only 0.1% survived 110 days, when their size was 3.45–15.9mm. However, no successful hatchery technique has yet been developed for *C opercularis,* many abnormal larvae being produced (Sasaki, 1979).

The spat of *Pecten maximus* has been hard to find in nature, being much scarcer than that of *Chlamys opercularis.* Both species were found on a variety of substrata above the sea bed. In experiments in Loch Torridon, west Scotland, in 1967 and 1968 numbers of pectinid spat settled on artificial structures, such as Netlon and Courlene rope, mostly within 2m of the sea bed in 21m of water. Most were *C opercularis,* but some were *P maximus,* the first time that appreciable numbers of *P maximus* spat had been recorded (Mason, 1969). Subsequently settlement experiments have been carried out in various places, including the west of Scotland (Ventilla, 1977; Sasaki, 1979), the Isle of Man (Paul, 1978; Brand, Paul and Hoogesteger, 1980), the south of England and north Wales (Pickett, 1978), the Republic of Ireland (Minchin, 1975; Anon, 1979) and France (Buestel *et al,* 1978). The collectors, which were set at varying heights above the sea bed, were of various sizes [*eg* 45 x 20cm (Paul, 1978), 100 x 30cm (Buestel *et al,* 1978), 100 x 50cm (Ventilla, 1977)], but basically of the same Japanese type of design. They consisted of Netlon, monofilament nylon or plastic mesh (*eg* onion bags) on which settlement occurred, within small mesh bags (approximately 2mm mesh) which ensured that the spat did not escape if it broke its byssus and left the collector (*Plate 22, Fig 37*). Spat collection has been highly successful in most years, though as in the initial experiments queens generally outnumbered scallops. For instance Ventilla (1977) obtained 85,000 queen and 10,000 scallop spat from 280 collectors in 1975 and 40,000 queen and 25,000 scallop spat from 110 collectors in 1976. In 1978 in Mulroy Bay, Co. Donegal, the settlement of *Pecten maximus* spat ranged from 10,000 to 60,000 per collector (Anon, 1979).

The best collections were made nearest the sea bed. Since settlements were of both species, if the species were to be grown on separately the spat would have to be separated manually. Because scallop spat is delicate, with a thin shell, this could result in damage. Paul (1978) found that the greatest settlement of *Pecten maximus* was nearer to the sea bed (3–7m) than that of *Chlamys opercularis* (4–10m). Ventilla (1977) found that the

Plate 22 String of tetrahedral spat collectors

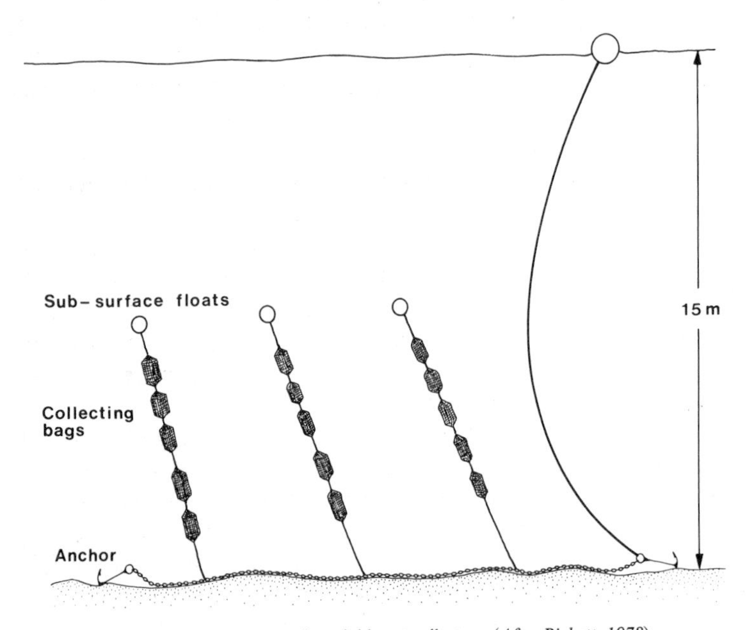

Fig 37 Arrangement of pectinid spat collectors. (*After Pickett, 1978*)

peak settlement periods of the two species were slightly different but overlapped. Further work might reveal that by putting out the collectors at appropriate depths at appropriate times it is possible to obtain settlements predominantly of one species or the other. In Japan, where there is an important industry based on collecting the spat of the related *Patinopecten yessoensis* (Ventilla, 1982), the plankton is monitored to determine when larvae are approaching the settlement time so that collectors can be set.

Growing spat in British waters has met with mixed success. Queens have been shown to grow well in suspended cages or bags, where they are protected from predators. For instance Mason and Drinkwater (1978) and Ventilla (1977) grew them to a mean size of 62mm, a good commercial size, in 1½–2 years from settlement in two Scottish west coast lochs, and mortalities were negligible. Scallops, however, have met with less success. Survival in hanging culture depends on the size at which spat is removed from the collectors (Buestel *et al,* 1978). After nine months survival of spat transferred at 3mm was only 40% compared with 85% in spat put out at 8mm. Mortalities are high in the first fortnight after transfer owing to the fragility of the scallop spat. Initial growth in suspension has been good. For example, Mason and Drinkwater (1978) grew them in suspension to lengths of up to 51mm within one year of settlement, but growth in the second year was poor, the greatest size achieved being 57mm. In cages on the sea bed in Loch Torridon, scallops attained an average length of only 75mm in four years compared with 100–110mm attained in nature in Scottish waters (J Mason, unpublished data). The poor growth might be attributed to nearness to the surface of the water (2–5m) since the scallops became heavily coated with tube worms which would hamper growth. Ventilla (1977) obtained a size of 35mm in 18 months, and suggested that with better growth at 20m than at 5m, 100–120mm could be expected in four to five years on the west coast of Scotland, though this seems optimistic.

Various methods of growing young scallops have been tried, including suspension in plastic mesh trays or Japanese-type lantern nets (*Plate 23*) (*eg* Ventilla, 1977), the latter being the more successful. Minchin (1975) attached young scallops to hanging ropes by means of Courlene thread through holes in the auricle, but these grew poorly and heavy fouling of the shell occurred. Fouling by epizoons would be less likely to occur if the scallops

Plate 23 Japanese-type lantern nets containing scallops

were put out on the natural substratum where they could recess.

Recent French work at St Brieuc and Brest has given hope for the success of scallop cultivation in British waters, together, possibly with the restocking of formerly productive grounds or even the stocking of grounds which appear suitable for scallops but where they do not settle abundantly. Spat of *Pecten maximus* were taken from collectors in St Brieuc Bay in September 1976 when, with an average size of 16mm, they were too small to be released onto the sea bed and were kept in broad baskets. Numbers were transferred under humid conditions to Brest, a journey of three hours, and fewer than 1% died. In March 1977, when their average size was almost 26mm, they were scattered on a former oyster bed in the Rade de Brest which had been levelled and cleared of starfish. Survival was good, about 50% being alive 15 months later in June 1978. Growth was excellent, reaching 85mm breadth (95mm length), almost commercial size, in late August 1978, 18 months after sowing and a little over two years after settlement (Buestel and Dao, 1978). Sowing was at a denisty of one scallop per $2m^2$, and future experiments are intended with densities up to five per m^2 as in the Japanese culture of *Patinopecten yessoensis*.

The advantages and drawbacks of hanging culture and restocking are illustrated by the Japanese industry along the coast of the Okhotsk sea in Hokkaido (Uno and Walford, 1977; Ventilla, 1982). There was a fishery based on a natural population, and in addition hanging culture was carried on using natural spat from collectors. Owing to overexploitation the yield from the fishery fell steadily to only 100t per year in the late 1960s. The spat taken for suspended culture to commercial size far exceeded the needs, and the excess were used to restock the natural beds. Restocking started in 1971 and 60 million seed 30mm long were released each year, harvesting starting in 1974. In 1974 1,700t were taken and in 1975 4,200t, with the remaining stock estimated at 8,000t. By 1977 the population had reached the same level as 30 years earlier and was increasing as a result of enhanced natural recruitment as well as restocking. The two fundamental requirements were removal of predators (starfish were killed by dredging or lime) and a stocking density of five or six scallops per m^2. Maintaining the optimum level of production has proved a problem. Over-production has occurred, resulting in a build up of bio-deposits and pollution, causing mass mortality. Natural food is limited and has proved insufficient, resulting in poor growth and quality. In the most heavily cultivated areas scallop densities on the sea bed have been as high as 40 per m^2, in addition to those in hanging culture in mid-water. This is a rare example of mariculture technology being too effective, and attempts are now being made to establish optimum levels for both hanging culture and restocking.

The success of the Japanese hanging scallop culture has prompted research into its feasibility in many parts of the British Isles. Results from experimental rearing of queens from natural spat have been very encouraging and, despite the low price, the abundance of spat, its ease of handling and good growth and survival have suggested that commercial scale tests would be worthwhile. Scallops in suspension have fared less well, but further research into the best conditions for growth and survival may well open the way for commercial operations. The sheltered pest-free inlets of western Scotland and Ireland should be ideal places for them.

The highly successful restocking of depleted Japanese scallop beds prompted the promising French experimental relaying of *Pecten maximus*. Similar research is now under way in the British Isles and again sheltered inlets where the sea bed is suitable appear to offer the best prospects for restocking old beds or even stocking new ones. There is more doubt about the feasibility of restocking or stocking with queens owing to their low price and greater mobility.

One final word on the legal aspects of cultivating and restocking. In Japan, farmed fish and shellfish production in coastal areas has priority over virtually all other users, be they navigational, recreational or industrial, and considerable co-operation takes place on the planned use of a coastal area for cultivation (Mackenzie and Johnston, 1976). The same is not true at least of the United Kingdom where, though rights may be granted by the Crown for particular areas of sea for mollusc cultivation, and the shellfish are the property of the lessee, rights of navigation must be preserved. In a restocked fishery there would be no private rights, since fishing on such beds would be a public right. The question arises as to who would foot the bill for such a practice if it proved feasible, the fishermen who benefit (in Japan local co-operatives have rights to their own beds), or the Government or one of its agencies. The legal position of scallops laid in new areas is uncertain, and would have to be clarified before cultivation, restocking and stocking of scallops could be undertaken commercially.

Perhaps the whole matter of scallop cultivation can best be summed up in the words of Motoda (1977) – 'Mariculture is an economic undertaking, but from the viewpoint of biological economy it is most rational to produce tasty animal protein by utilizing natural primary production'. There is a virtually insatiable demand for scallops and queens in Europe and America, which could be increased by market promotion. Since both are luxury species, the natural fishery, with appropriate conservation and possibly augmented by cultivation and restocking, should have a healthy future.

Bibliography

Amirthalingam, C., 1928. On lunar periodicity in reproduction of *Pecten opercularis* near Plymouth in 1927–28. *J. mar. biol. Ass. U.K.,* Vol 15, pp 605–641.

Andreu, B., 1968. The importance and possibilities of mussel culture. Working Paper 5. Symposium on possibilities and problems of fisheries development in south-east Asia, Berlin (Tegel), 10–30 September 1968, pp 364–375. Berlin, German Foundation for Developing Countries.

Anon, 1978. Triennial review of research at the Marine Laboratory, Aberdeen, for the years 1973–1975. Edinburgh, Department of Agriculture and Fisheries for Scotland, 85pp.

Anon, 1979. Escallop scheme proves a success. *The Irish Skipper,* No. 190, November 1979, p8.

Antoine, L., 1978. La croissance journalière chez *Pecten maximus.* Proceedings of the 2nd Pectinid Workshop, Brest, May 1978, 2pp.

Aravindakshan, I., 1955. Studies on the biology of the queen scallop *Chlamys opercularis* (L.). University of Liverpool, Ph.D. Thesis, 79 pp.

Askew, C.G., Dunn, M.R. and Reay, P.J., 1974. The fishery for queen scallops in Guernsey. Marine Resources Research Unit, Portsmouth Polytechnic, 38pp (mimeo).

Baird, R.H., 1952. The English Channel escallop beds. *Fishing News,* No. 2064, 8 November 1952, p9.

Baird, R.H., 1955. A preliminary report on a new type of commercial escallop dredge. *J. Cons. perm. int. Explor. Mer,* Vol. 20, pp 290–294.

Baird, R.H. 1958. On the swimming behaviour of escallops (*Pecten maximus* L.). *Proc. malac. Soc. Lond.*, Vol. 33, pp 67–71.

Baird. R.H., 1959. Factors affecting the efficiency of dredges. H. Kristjónsson, ed. *Modern fishing gear of the world.* London, Fishing News (Books) Ltd, pp 222–224.

Baird, R.H., 1966. Notes on an escallop (*Pecten maximus*) population in Holyhead harbour. *J. mar. biol. Ass. U.K.*, Vol. 46, pp 33–47.

Baird, R.H., and Gibson, F.A., 1956. Underwater investigations on escallop (*Pecten maximus* L.) beds. *J. mar. biol. Ass. U.K.,* Vol. 35, pp 555–562.

Bellew, G., 1957. Escallops in armory. *Cox, I. ed. The Scallop.* London, Shell Transport and Trading Company Ltd, pp 89–104.

Bhatnagar, K.M., 1972. East coast queen fishery 1970. Fisheries Leaflet No. 32, Dublin. 5pp.

Brand, A.R., Paul, J.D. and Hoogesteger, J.N., 1980. Spat settlement of the scallops *Chlamys opercularis* (L.) and *Pecten maximus* (L.) on artificial collectors. *J. mar. biol. Ass. U.K.,* Vol. 60, pp 379–390.

Broom, M.J., 1976. Synopsis of biological data on scallops *Chlamys (Aequipecten) opercularis* (Linnaeus), *Argopecten irradians* (Lamarck), *Argopecten gibbus* (Linnaeus). FAO Fish. Synops., No 114, 44pp.

Broom, M.J. and Mason, J., 1978. Growth and spawning in the pectinid *Chlamys opercularis* in relation to temperature and phytoplankton concentration. *Mar. Biol.,* Vol. 47, pp 277–285.

Buddenbrock, W. von and Möller-Racke, I., 1953. Über den Lichtsinn von *Pecten. Pubbl. zool. Staz. Napoli,* Vol. 24 pp 217–245.

Buestel, D., Arzel, P., Cornillet, P. and Dao, J. – C., 1978. Production de juveniles de coquilles Saint-Jacques (*Pecten maximus* (L.)). *Actes de Colloques du CNEXO,* Vol. 4 , pp 307–315.

Buestel, D. and Dao, J. – C., 1978. Aquaculture extensive de la coquille St Jacques: resultats d'un semis experimental. Contrib. No. 624, Départment scientifique du Centre oceanologique de Brest. Mimeo. 10pp.

Buestel, D., Dao, J. – C. and Lemarié, G., 1979. Collecte du naissain de pectinidés en Bretagne. *Rapp. P. – V. Réun. Cons. int. Explor. Mer,* Vol. 175, pp 80–84.

Chapman, C.J., Main, J., Howell, T. and Sangster, G.I., 1979. The swimming speed and endurance of the queen scallop *Chlamys opercularis* in relation to trawling. *Progress in Underwater science,* Vol. 4, J.C. Gamble (ed.). J.D. George, Pentech Press, pp 57–72.

Chapman, C.J., Mason, J. and Kinnear, J.A.M., 1977. Diving observations of the efficiency of dredges used in the Scottish fishery for the scallop, *Pecten maximus* (L.). *Scott. Fish. Res. Rep.,* No 10, 16pp.

Clark, G.R. II, 1968. Mollusk shell: daily growth lines. *Science,* Vol. 161, pp 800–802.

Coe, W.R., 1933. Sexual phases in *Teredo. Biol. Bull., Woods Hole,* Vol. 65, pp 283–303.

Coe, W.R., 1945. Development of the reproductive system and variations in sexuality in *Pecten* and other pelecypod molluscs. *Trans. Conn. Acad. Arts. Sci.,* Vol. 36, pp 673–700.

Comely, C.A., 1972. Larval culture of the scallop *Pecten maximus* (L.) *J. Cons. int. Explor. Mer,* Vol. 34, No. 3, pp 365–378.

Comely, C.A., 1974. Seasonal variations in the flesh weights and biochemical content of the scallop *Pecten maximus* L. in the Clyde Sea Area. *J Cons. int. Explor. Mer,* Vol. 35, No. 3, pp 281–295.

Connor, P.M., 1978. Seasonal variation in meat yield of scallops (*Pectin maximus*) (*sic*) from the south coast (Newhaven) of England. ICES, CM 1978, Doc. No. K: 8, 2pp (Mimeo).

Cox, I., (ed), 1957. *The scallop.* London, Shell Transport and Trading Co. Ltd, 135pp.

Dakin, W.J., 1909. *Pecten. Mem. Lpool mar. biol. Comm.,* No. 17, 136pp.

Davis, F.M., 1927. An account of the fishing gear of England and Wales (revised edition). *Fish. Invest. Lond.,* Ser. II, Vol. 9, No. 6, 131pp.

Drinkwater, J., 1974. Scallop dredge selectivity experiments. ICES, CM 1974, Doc. No. K:25, 4pp (mimeo).

Duff, M., 1976. Scallop fishing in Ireland. Proceedings of the 1st Pectinid Workshop, Baltimore, May 1976, 9pp.

Early, J.C., and Stroud, G.D., 1981. Shellfish processing in Scotland. *Scott. Fish. Bull.,* No. 46, pp 29–35.

Eggleston, D., 1962. Spat of the scallop (*Pecten maximus* L.) from off Port Erin, Isle of Man. *Rep. mar. biol. Sta. Port Erin,* No. 74, pp 29–32.

Elmhirst, R., 1945. Clam fishing in the Firth of Clyde. *Trans. Butesh. nat. Hist. Soc.,* Vol. 13, pp 113–116.

Forbes, E. and Hanley, S., 1853. A history of British Mollusca and their shells. London, Van Voorst, 4 Volumes.

Franklin, A., Pickett, G.D., and Connor, P.M., 1980. The escallop (*Pecten maximus*) and its fishery in England and Wales. *MAFF Laboratory Leaflet,* No. 51, 19pp.

Fullarton, J.H., 1890. On the development of the common scallop (*Pecten opercularis* L.). *Rep. Fish. Bd Scot.,* No. 8, Pt 3, pp 290–299.

Gibson, F.A., 1956. Escallops (*Pecten maximus* L.) in Irish waters. *Sci. Proc. Roy. Dublin Soc.,* Vol. 27, pp 253–271.

Gibson, F.A., 1957. Escallop fishing around Ireland. *Rep. Sea. Inl. Fish., Dept. Lands Fish. Div.,* 1957, App. 24, pp 60–65.

Goodlad, M.H., 1976. Tests on a rotary sorter for queen scallops. *Scott. Fish. Bull,* No. 43, pp 44–46.

Gruffydd, Ll.D., 1972. Mortality of scallops on a Manx scallop bed due to fishing. *J. mar. biol. Ass. U.K.,* Vol. 52, pp 449–455.

Gruffydd, Ll.D., 1974a. The influence of certain environmental factors on the maximum length of the scallop, *Pecten maximus* L. *Cons. int. Explor. Mer,*

Vol. 35, No. 3, pp 300–302.

Gruffydd, Ll.D., 1974b. An estimate of natural mortality in an unfished population of the scallop *Pecten maximus* (L.). *J. Cons. int. Explor. Mer,* Vol. 35, No. 2, pp 209–210.

Gruffydd, Ll.D., 1976. Swimming in *Chlamys islandica* in relation to current speed and an investigation of hydrodynamic lift in this and other scallops. *Norw. J. Zool.,* Vol. 24, pp 365–378.

Gruffydd, Ll.D., 1981. Observations on the rate of production of external ridges on the shell of *Pecten maximus* in the laboratory. *J. mar. biol. Ass. U.K.,* Vol. 61, pp 401–411.

Gruffydd, Ll.D., and Beaumont, A.R., 1972. A method for rearing *Pecten maximus* larvae in the laboratory. *Mar. Biol.,* Vol. 15, pp 350–355.

Hancock, D.A., 1979. Population dynamics and management of shellfish stocks. *Rapp. P.–V. Réun. Cons. int. Explor. Mer,* Vol. 175, pp 8 – 19.

Hardy, D., 1981. *Scallops and the diver-fisherman.* Farnham, Fishing News Books Ltd, 134pp.

Hardy, R. and Smith, J.G.M., 1970. Catching and processing scallops and queens. *Torry Advisory Note,* No. 46, 11pp.

Hartnoll, R.G., 1967. An investigation of the movement of the scallop, *Pecten maximus. Helgol. wiss. Meeresunters.,* Vol. 15, pp 523–533.

Iversen, E.S., 1968. *Farming the edge of the sea.* London, Fishing News Books Ltd, 301pp.

Jørgensen, C.B., 1946. Lamellibranchia. *In* Thorson, G. Reproduction and larval development of Danish marine bottom invertebrates, with special reference to the planktonic larvae in the sound (Øresund). *Meddr Kommn Danm. Fisk. – og Havunders., Ser. Plankton,* Vol. 4, pp 277–311.

Kinne, O., 1970. Closing address of international symposium on cultivation of marine organisms and its importance for marine biology. *Helgol. wiss. Meeresunters.,* Vol. 20, pp 707–710.

Lacaille, A.D., 1954. *The stone age in Scotland.* London, Oxford University Press, 345pp.

Land, M.F., 1966. Activity in the optic nerve of *Pecten maximus* in response to changes in light intensity and to pattern and movement in the optical environment. *J. exp. Biol.,* Vol. 45, pp 83–99.

Lovell, M.S., 1884. *The edible Mollusca of Great Britain and Ireland* (2nd Edn). London, Reeve, 310pp.

Mackenzie, W.D., and Johnston, A.J.A., 1976. Visits to various fish farming units in Japan, May – June 1976. Inverness, Highlands and Islands Development Board, 23pp (mimeo).

Mason, J., 1957. The age and growth of the scallop, *Pecten maximus* (L.), in

Manx waters. *J. mar. biol. Ass. U.K.*, Vol. 36, pp 473–492.
Mason, J., 1958a. The breeding of the scallop, *Pecten maximus* (L.), in Manx waters. *J. mar. biol. Ass. U.K.*, Vol. 37, pp 653–671.
Mason, J., 1958b. A possible lunar periodicity in the breeding of the scallop, *Pecten maximus* (L.). *Ann. Mag. nat. Hist.*, Ser. 13, Vol. 1, pp 601–602.
Mason. J., 1959a. The state of Manx scallop stocks 1950–53. *Rep. mar. biol. Sta. Port Erin,* No. 71, pp 39–46.
Mason, J., 1959b. The food value of the scallop, *Pecten maximus* (L.), from Manx inshore waters. *Rep. mar. biol. Sta. Port Erin,* No. 71, pp 47–52.
Mason, J., 1969. The growth of spat of *Pecten maximus* (L.) ICES, CM 1969, Doc. No. K:32, 3pp (mimeo).
Mason, J., 1970. A comparison of various gears used in catching queens and scallops in Scottish waters. ICES, CM1970, Doc. No. K:19, 3pp (mimeo).
Mason, J., 1972a. The Scottish fishery for scallops and queens. *Scott. Fish. Info. Pamphl.* Aberdeen, Marine Laboratory, 11pp.
Mason, J., 1972b. The cultivation of the European mussel, *Mytilus edulis* Linnaeus. *Oceanogr. mar. Biol. Ann. Rev.*, Vol. 10, pp 437–460.
Mason, J., 1976. Cultivation. *Bayne, B.L., ed. Marine Mussels.* Cambridge, University Press, pp 385–410.
Mason, J., 1978. The Scottish scallop fishery. *Scott. Fish. Bull.,* No. 44, pp 38–44.
Mason, J., 1980. The Scottish fishery for the queen, *Chlamys opercularis* (L.) Proceedings of the 3rd Pectinid Workshop, Port Erin, May 1980, 6pp.
Mason, J., Chapman, C.J. and Kinnear, J.A.M., 1979. Population abundance and dredge efficiency studies on the scallop, *Pecten maximus* (L.). *Rapp. P.–V. Réun. Cons. int. Explor. Mer,* pp 91–96.
Mason, J. and Colman, J.S., 1955. Note on a short-term marking experiment on the scallop *Pecten maximus* (L.) in the Isle of Man. *Rep. mar. biol. Sta. Port Erin,* No. 67, pp 34–35.
Mason, J. and Drinkwater, J., 1969. Scallops (*Pecten maximus* (L.)) in the Firth of Clyde. ICES, CM 1969, Doc. No. K:33, 3pp (mimeo).
Mason, J. and Drinkwater, J., 1973. The scallop fishery off south-west Scotland. *Scott. Fish. Bull.,* No. 39, pp 40–44.
Mason, J. and Drinkwater, J., 1974. The stocks of scallops (*Pecten maximus* (L.)) in the Clyde sea area and west of Kintyre. *Annls. biol., Copenh.,* Vol. 29, pp 184–186.
Mason, J. and Drinkwater, J., 1975. The stocks of scallops, *Pecten maximus,* in the Clyde sea area and west of Kintyre in 1972–73. *Annls. biol. Copenh.,* Vol. 30, pp 213–214.
Mason, J. and Drinkwater, J., 1976. The stocks of scallops, *Pecten maximus,*

in the Clyde sea area and west of Kintyre in 1973–74. *Annls. biol., Copenh.,* Vol. 31, pp 183–184.

Mason, J. and Drinkwater, J., 1978. The settlement and early growth of the scallop, *Pecten maximus* (L.), and the queen, *Chlamys opercularis* (L.), in Scottish waters. Proceedings of the 2nd Pectinid Workshop, Brest, May 1978, 4pp.

Mason, J., Nicholson, M.D., and Shanks, A.M., 1979. A comparison of exploited populations of the scallop, *Pecten maximus* (L.). *Rapp. P.–V. Réun. Cons. int. Explor. Mer,* Vol. 175, pp 114–120.

Mason, J., Shanks, A.M. and Fraser, D.I., 1980. An assessment of scallop, *Pecten maximus* (L.), stocks off south-west Scotland. ICES, CM 1980, Doc. No. K:27, 4pp (mimeo).

Mason, J., Shanks, A.M. and Fraser, D.I. 1981. An assessment of scallop *Pecten maximus* (L.) stocks at Shetland. ICES, CM 1981, Doc. No. K:19, 4pp (mimeo).

Mason, J., Shanks, A.M. Fraser, D.I. and Shelton, R.G.J., 1979. The Scottish fishery for the queen, *Chlamys opercularis* (L.) ICES, CM 1979, Doc. No. K:37, 6pp (mimeo).

Mathers, N.F., 1976. The effects of tidal currents on the rhythm of feeding and digestion in *Pecten maximus* L) *J. exp. mar. Biol. Ecol.,* Vol. 24, pp 271 –283.

Minchin, D., 1975. Experimental hanging culture of *Pecten maximus* in the West of Ireland, with a note on tagging. ICES, CM 1975, Doc. No. K:3, 5pp (mimeo).

Minchin, D., 1978a. The behaviour of young escallops (*Pecten maximus* L. –Pectinidae). Proceedings of the 2nd Pectinid Workshop, Brest, May 1978, 11pp.

Minchin, D., 1978b. An excehtionally large escallop (*Pecten maximus* (L.)) from West Cork. *Irish Nat. J.,* Vol. 19, p 202.

Minchin, D., 1980. Room for the scallop. *Fish Farmer,* Vol. 3 (5), pp 18–21.

Moore, J.D. and Trueman, E.R., 1971. Swimming of the scallop, *Chlamys opercularis* (L.). *J. exp. mar. Biol. Ecol.,* Vol. 6, pp 179–185.

Motoda, S., 1977. Biology and artificial propagation of Japanese scallop (general review). *Motoda, S., ed.* Proceedings of the *Second Soviet–Japan Joint Symposium on Aquaculture, November, 1973.* Tokyo, Tokai University, pp 75–120.

Orton, J.H., 1928. On rhythmic periods in shell growth in *O. edulis* with a note on fattening. *J. mar. biol. Ass. U.K.,* Vol. 15, pp 365–427.

Paul, J.D., 1978. The biology of the queen scallop, *Chlamys opercularis* (L.), in relation to its prospective cultivation. University of Liverpool, Ph.D. Thesis.

Pickett, G.D., 1978. The scallop. *Underwater World,* Vol. 1, No. 6, pp 14–16.

Pickett, G.D. and Franklin, A., 1975. The growth of queen scallops (*Chlamys opercularis*) in cages off Plymouth, south-west England. ICES, CM 1975, Doc. No. K:25, 4pp. (mimeo)

Pope, J.A. and Mason, J. 1980. The fitting of growth curves for *Pecten maximus* (L.). ICES, CM 1980, Doc. No. K:28, 5pp.

Rees, C.B., 1950. The identification and classification of lamellibranch larvae. *Hull Bull. mar Ecol.,* Vol. 3, No. 19, pp 73–104.

Rees, W.J., 1957. The living scallop. *Cox, I. ed. The Scallop.* London, Shell Transport and Trading Company Ltd, pp 15–32.

Ridgway, R., 1912. Color standards and color nomenclature. Published by the author. Washington, D.C.

Rolfe, M.S., 1969. The determination of the abundance of escallops and of the efficiency of the Baird escallop dredge. ICES, CM 1969, Doc. No. K:22, 5pp.

Rolfe, M.S., 1973. Notes on queen scallops and how to catch them. *MAFF Shellfish Information Leaflet,* No. 27, 13pp.

Sasaki, R., 1979. A report on the study of scallop and oyster in the course of Japan/Scotland exchange research scholarship 1977/1978. Inverness, Highlands and Islands Development Board, 23pp.

Soemodihardjo, S., 1974. Aspects of the biology of *Chlamys opercularis* (L.) (Bivalvia) with comparative notes on four allied species. University of Liverpool, Ph.D. Thesis, 110pp.

Stanley, C.A., 1967. The commercial scallop, *Pecten maximus* (L.), in Northern Irish waters. The Queen's University, Belfast, Ph.D. Thesis, 111pp.

Strange, E.S., 1977. An introduction to commercial fishing gear and methods used in Scotland. *Scott. Fish. Info. Pamphl.,* No. 1, 34pp.

Strange, E.S., 1979. Scallop dredging gear investigations; comparative fishing with experimental gear. Aberdeen, Marine Laboratory Working Paper No. 79/4, 5pp.

Tang, S.–F., 1941. The breeding of the escallop (*Pecten maximus* (L.)) with a note on the growth rate. *Proc. Lpool biol. Soc.,* Vol. 54, pp 9–28.

Taylor, A.C. and Venn, T.J., 1978. Growth of the queen scallop, *Chlamys opercularis,* from the Clyde Sea area. *J. mar. biol. Ass. U.K.,* Vol. 58, pp 687–700.

Taylor A.C. and Venn, T.J., 1979. Seasonal variation in weight and biochemical composition of the tissues of the queen scallop, *Chlamys opercularis,* from the Clyde Sea area. *J. mar. biol. Ass. U.K.,* Vol. 59, pp 605–621.

Tebble, N., 1966. *British bivalve seashells.* London, British Museum (Natural History), 212pp.

Thomas, G.E. and Gruffydd, Ll. D., 1971. The types of escape reactions elicited in the scallop *Pecten maximus* by selected sea-star species. *Mar. Biol.,* Vol. 10, pp 87–93.

Uno, Y. and Walford, J., 1977. Recent mariculture technology in Japan with special regard to abalone. Shellfish Association of Great Britain, 8th Shellfish Conference, May 1977, pp 1–11.

Ventilla, R.F., 1977. Preliminary investigations into the spat collection, nursery maintenance and on-growing of the scallops *Pecten maximus* and *Chlamys opercularis*. Shellfish Association of Great Britain, 8th Shellfish Conference, May 1977, pp 12–16.

Ventilla, R.F., 1982. The scallop industry in Japan. *Advances in Marine Biology,* Vol. 20, pp 309–382.

Index

Other books published by
Fishing News Books Ltd

Free catalogue available on request